ECOTOURISM AND ENVIRONMENTAL SUSTAINABILITY

To our mothers, Janet Hill and Mercia Gale,
for their unfailing love, guidance and friendship.

Ecotourism and Environmental Sustainability

Principles and Practice

Edited by
JENNIFER HILL AND TIM GALE
University of the West of England, UK

ASHGATE

Published by
Ashgate Publishing Limited
Wey Court East
Union Road
Farnham
Surrey, GU9 7PT
England

Ashgate Publishing Company
Suite 420
101 Cherry Street
Burlington
VT 05401-4405
USA

www.ashgate.com

British Library Cataloguing in Publication Data
Ecotourism and environmental sustainability : principles and practice
1. Ecotourism 2. Environmental protection 3. Environmental degradation 4. Ecotourism - Case studies 5. Environmental protection - Case studies 6. Environmental degradation - Case studies
I. Hill, Jennifer (Jennifer Louise), 1969- II. Gale, Tim
338.4'791

Library of Congress Cataloging-in-Publication Data
Ecotourism and environmental sustainability : principles and practice / by Jennifer Hill and Tim Gale.
p. cm.
Includes index.
ISBN 978-0-7546-7262-3 -- ISBN 978-0-7546-7621-8 (ebook) 1. Ecotourism--Environmental aspects. 2. Ecotourism. I. Hill, Jennifer (Jennifer Louise), 1969- II. Gale, Tim.

G156.5.E26E353 2009
910.68'4--dc22

2009002110

ISBN 978 0 7546 7262 3 (hardback)
eISBN 978 0 7546 7621 8 (ebook)

Mixed Sources
Product group from well-managed forests and other controlled sources
www.fsc.org Cert no. SA-COC-1565
© 1996 Forest Stewardship Council
FSC

Printed and bound in Great Britain by
MPG Books Ltd, Bodmin, Cornwall.

Contents

List of Figures

List of Tables

List of Contributors

Georgette Leah Burns is a lecturer at Griffith University (Queensland, Australia), in the Griffith School of Environment, and she is a foundation member of the Centre for Innovative Conservation Strategies. She has undertaken extensive fieldwork in Nepal, Fiji and Australia, studying the impacts of tourism on local communities. Her current research, in the field of environmental anthropology, focuses on the interactions between people and wildlife in wildlife tourism settings. Her work has appeared in journals (including *Tourism Management*, *Anthropological Forum*, *Australasian Journal of Environmental Management*), as well as book chapters (in *Pacific Tourism*, *Wildlife Tourism*), and it has received media coverage. She is currently preparing a manuscript for a book on wildlife tourism case studies.

Alton C. Byers is a mountain geographer, climber and photographer, specializing in high altitude conservation programmes, applied research and climate change impacts in the mountains. He received his doctorate from the University of Colorado in 1987, focusing on landscape change, soil erosion and vegetation dynamics in the Sagarmatha (Mt. Everest) National Park, Khumbu, Nepal. He has worked for The Mountain Institute in West Virginia, USA, since 1990 in its Asian, Andean, North American and African Programmes and has published widely on a variety of field research topics. His recent awards include the Nature Conservancy's 'Mountain Stewardship Award'; the American Alpine Club's 'David Brower Conservation Award'; the Association of American Geographer's 'Distinguished Career Award'; and the 'Sir Edmund Hillary Mountain Legacy Medal' for 'distinguished service to mountain people and environments in remote regions'. Since 2007 he has directed the Alpine Conservation Partnership project that, in collaboration with the American Alpine Club, is working toward the conservation and restoration of fragile alpine ecosystems worldwide.

Terry D. DeBruyn is the Regional Wildlife Biologist for the National Park Service's Alaska Region. He received a BS in Environmental Science from Grand Valley University (1976), and both a MS in Biology, from Northern Michigan University (1992), and a PhD from Michigan Technological University (1997) studying black bears in Michigan's Upper Peninsula. Among his other experiences with bears are Section Leader, Bear Research and Management, Florida Conservation Commission and Bear Biologist, Katmai National Park and Preserve, Alaska. Terry is primarily interested in field research of bear habitat use, movement and behaviour. He has studied bears for the past 18 years and published a popular book *Walking with Bears* about his years of bear study in the Upper Peninsula of Michigan.

Dianne Draper is a Professor in the Department of Geography, University of Calgary, and has authored numerous publications on environmental management, sustainable tourism and coastal zone management. Her research focuses on governance and quality of life issues in communities working toward sustainable tourism, and on managing tourism growth and its impacts on communities, water resources, and parks and protected areas.

Simon Evans is a Senior Lecturer in Sustainable Tourism and a researcher within the Tourism, Sustainability, Leisure and Aviation Research Group of the Ashcroft International Business School at Anglia Ruskin University, UK. His research interests and expertise are in ecotourism, sustainable tourism practices and policies, and wildlife tourism.

Tim Gale is a Senior Lecturer in Tourism Geography and Programme Leader for the BA(Hons) Tourism Management/and Environmental Management programme(s) at the University of the West of England, Bristol. His research interests include the relationship between late twentieth century processes of economic restructuring and cultural change and the built environment of cold-water resorts, new forms of tourism production and consumption (e.g. cyber- and eco-tourism), and environmental sustainability and tourism development. He is particularly interested in the potential for new knowledge creation, with regards to each of these areas of enquiry, offered by the so-called 'mobilities turn' (integrating leisure and tourism, transport, and migration studies), and critical realism as a philosophy of/for the social sciences. He has authored or co-authored a number of journal articles, book chapters and conference papers to date that contribute to various discourses in tourism studies (e.g. the applications and limitations of the tourist area life cycle, 'the end of tourism'), which are supported by a range of empirical materials (e.g. local authority committee minutes, rateable value statistics, photographs within holiday brochures). Ongoing projects include completing a survey of visitors to Bristol Urban Beach 2007, and evaluating alternative delivery methods in undergraduate tourism studies.

Georgie Gough is a geographer whose research interests cover many aspects of the nature-based tourism industry. Her research has focused on interpretation provision at nature-based visitor attractions, especially within wetland and rainforest environments. She has published in the *Journal of Ecotourism*, in addition to producing book chapters and conference presentations.

Jennifer Hill is a Principal Lecturer in Geography and Environmental Management at the University of the West of England, Bristol, UK. Her research interests comprise diverse yet inter-related themes within environmental resource management. She has studied the biodiversity and conservation of tropical forests in Ghana, biodiversity interpretation in the Australian rainforest, the sustainable management of bio-resources through community-based ecotourism in Peru, and resource use by societies in semi-arid Tunisia. Jenny has authored/co-authored 21 journal articles (published in, amongst others, *Journal of Ecotourism*, *Geographical Journal*, *Journal of Biogeography*, *Progress in Physical Geography*, *Applied Geography*, and *Global Ecology and Biogeography*). She has published 40 other articles including conference papers and abstracts, invited presentations in the education and public sectors, book chapters and consultancy reports. Additionally, Jenny is lead editor/co-editor of 2 published books: *Making Space: Managing Resources for Leisure and Tourism* (Leisure Studies Association, 2007) and *Sustainable Development: National Aspirations, Local Implementation* (Ashgate Publishing Limited, 2006).

Andrew Holden is Professor of Environment and Tourism and also the Director for the Centre for Research into the Environment and Sustainable Tourism Development (CREST) at the University of Bedfordshire in England. He has written widely on the themes of the inter-relationship between tourism, environment and development. He sits on the editorial boards of three academic journals, including *Annals of Tourism Research*. Invited key note talks at international conferences have included the United Nations World Tourism Organisation in Madrid; Arctic Research Institute in Finland; the World Wide Fund for Nature in Athens; and the European Association for Tourism and Leisure (ATLAS) in Portugal. His past employment has included acting as the Head of Research for the Faculty of Leisure and Tourism at Buckinghamshire Chilterns University College and he has held research fellow positions in several British universities. He has also previously worked in the private sector, for the Economic Studies Group, as a consultant for tourism development projects. His research and consultancy work has been undertaken in Indonesia, Nepal, Turkey, Russia and Cyprus, alongside studies in the United Kingdom.

Colin Hunter is Reader in Geography and Environment in the School of Geosciences at the University of Aberdeen. He has been researching aspects of environmental management and sustainable development for approaching 20 years, has co-authored two books and published some 40 research papers and book chapters across a wide range of topics, including sustainable tourism, water resources management and household environmental impact. Much of his research activity bridges the natural and social science traditions, most recently extending to encompass interests in disease risk perception and the use of e-Social Science computing technologies to promote interdisciplinary collaboration in sustainability research. He is currently working on research projects worth over £3 million, funded by UK Research Councils, the Scottish Government and the European Union.

Wenjun Li is Professor of Environment in the College of Environmental Sciences and Engineering at Peking University. Her research focuses on community-based management of natural resources, public administration of protected areas, and property rights and rangeland management in arid and semi-arid grassland. She has published on natural resource management in a range of journals, including *Journal of Environmental Management*, *Environmental Management*, *AMBIO*, *Annals of Tourism Research*, *Environmental Monitoring and Assessment*, *Tourism Management*, *and Biological Conservation*.

Ken Simpson is director of postgraduate business studies programmes at Unitec Institute of Technology in Auckland, New Zealand. His past research and teaching interests have been based on regional- and local-level strategic planning for future tourism developments, and previous publications have examined the attitudes and behaviours of both tourism operators and tourists (notably 'wine tourists') in rural destinations. He is currently engaged in a multi-national research initiative that seeks to develop improved understanding of the economic impacts of non-resident spending in 'transit zone' towns, those settlements that are situated on surface travel corridors that are widely used by travellers en route to popular tourist destinations.

Tom S. Smith is an Associate Professor and Research Wildlife Biologist in the Wildlife Sciences Programme, Faculty of Plant and Animal Sciences, Brigham Young University, Provo, UT. Tom earned a BS in Zoology from Brigham Young University in 1982, a MS in Wildlife Management at the University of Alaska in 1987, and a PhD in Wildlife Ecology at Brigham Young University in 1992. Tom has studied bear–human conflict for the past 16 years and maintains a database of over 550 bear attacks and has studied them extensively. He currently studies polar bears with a focus on den emergence ecology and response to human activities on the North Slope, Alaska.

Davina Stanford currently works as a Research Associate for TEAM Tourism Consulting. Amongst many projects with TEAM, Davina has recently co-ordinated, researched and contributed to a UNWTO publication *A Practical Guide to Tourism Destination Management* (an authoritative reference handbook concerning the full spectrum of activities involved in integrated destination management). Her most extensive research has concentrated on understanding the role of the tourist in achieving responsible tourism and she has published on this subject in the *Journal of Sustainable Tourism*.

Emma J. Stewart, originally from the UK, completed her Geography PhD at the University of Calgary in 2008. She holds a Trudeau Foundation Doctoral Scholarship to support her research which explores resident attitudes toward tourism development in the Arctic. This builds on research experience in Antarctica and New Zealand where she completed her Masters degree at Lincoln University.

Lucy A. Sutherland is a Research Officer with the University of Queensland's School of Tourism in Australia. Lucy began her career working in botanic garden and protected area management in Australia and has more recently worked internationally. She uses this field experience to support her academic research in the use of modified spaces for ecotourism and its contribution to conservation and sustainable development. Lucy has a particular interest in tourism as a sustainable use of biodiversity and was co-author and principal editor of the *International Agenda for Botanic Gardens in Conservation*, which was adopted by the Secretariat for the Convention on Biological Diversity as a contribution by the world botanic gardens to the *Global Strategy for Plant Conservation*.

Preface

This edited collection has its origins in the Royal Geographical Society (with the Institute of British Geographers) (RGS-IBG) Annual International Conference 2006, held at the Society's headquarters in Kensington Gore, London. At this conference the Biogeography Research Group and the Geography of Leisure and Tourism Research Group of the RGS-IBG co-sponsored a session entitled 'Ecotourism and Environmental Sustainability'. We convened this session, with the intention of interrogating the complex interactions between ecotourism and a range of 'natural' environments, from polar to tropical and terrestrial to aquatic. Notionally, ecotourism helps to conserve these environments, many of which could be construed as 'fragile', in addition to helping improve the welfare of local people (thus embodying the principles of inter- and intra-generational equity that lie at the heart of sustainable development); it can also engender more positive attitudes towards the environment and desirable behaviour in participants, both at the destination and following their return home. However, ecotourism has also attracted criticism, not least with regards to the environmental cost of air travel to popular destinations such as Costa Rica and Kenya, the 'true' motivations of ecotourists (which appear to have as much to do with sustaining the ego as the environment), and the potential for unwittingly disturbing ecosystems through co-presence.

With this in mind, we invited papers from either side of the natural–/social–scientific divide that offered theoretical or practical insights into the complex interactions between ecotourism and the natural environment, from the local to the global scale. This book incorporates papers presented at the aforementioned RGS-IBG conference, together with contributions from selected academics and practitioners. The volume examines critically the ambivalence that lies behind the concept of ecotourism as synonymous with environmentally sustainable tourism; likewise, it resists a dominant trend to showcase best (or most sustainable) practice.

We would like to take this opportunity to thank all contributors for their timely return of draft and final manuscripts, and for their sustained engagement with a project of some length. We would also like to thank Paul Revell from the Cartographic Unit at the University of the West of England, Bristol for production of many of the figures that appear in the book. Our final thanks are extended to our colleagues, students and loved ones who had to bear with us as we buried ourselves under proofreading, organization of content and synthesis of ideas, all of which distracted us from myriad other duties. Their forbearance and understanding has ultimately allowed this book to come to print.

Jennifer Hill and Tim Gale
2009

PART 1
The Context of Ecotourism and Environmental Sustainability

Chapter 1
Ecotourism and Environmental Sustainability: An Introduction

Tim Gale and Jennifer Hill

Ecotourism: Its Meaning and Significance

Thanks to the creation of previously inaccessible and undeveloped destinations, and a preference for independent and special interest holidays in non-resort locations, there is little left of the natural world that has not been exploited or commoditized for tourist consumption. The extension of the so-called 'pleasure periphery' (after Turner and Ash 1975; see Figure 1.1) into ever more remote and exotic areas has been driven to a significant degree by various forms of nature-based tourism, especially ecotourism, which are manifest in a range of environments from polar to tropical and terrestrial to aquatic, and which exhibit a strong correlation with periphernortal and (predominantly public) protected areas. Nature-based tourism in general is one of the fastest growing sectors within the global tourism industry (Buckley 2000, Ryan et al. 2000, Wight 2001, Kuo 2002), it being defined as tourism 'primarily concerned with the direct enjoyment of some relatively undisturbed phenomenon of nature' (Valentine 1992: 108). Market estimates are hard to come by, given the lack of consensus over the use of the term(s), but it was suggested in 2004 that eco-/nature tourism was growing three times faster globally than the tourism industry as a whole (WTO 2004, cited in TIES 2006). Reasons for this growth include demographic changes in source countries (such as older populations and, in turn, the growing number of more experienced travellers), 'beach boredom' as a symptom of a maturing market for 3S (sun, sea, sand) holidays and increasing environmental awareness on the part of the general public (Ayala 1996). It is also attributable to the rapid development of an industry comprised of specialized (for example ecolodges, ecotour operators, and suppliers of transport services and infrastructures within a given ecotourism destination) and non-specialized businesses (such as hotel chains, airline and cruise ship operations, and retail travel agents), ranging from small- and medium-sized enterprises to transnational corporations (Weaver and Lawton 2007).

Under the broad banner of nature-based tourism, ecotourism has been suggested as a way in which increasing numbers of visitors seeking an intrinsically environmental tourism experience can be accommodated, whilst minimizing the costs and enhancing the benefits associated with natural area tourism (Boo 1990, Cater and Lowman 1994). Many definitions and types of ecotourism

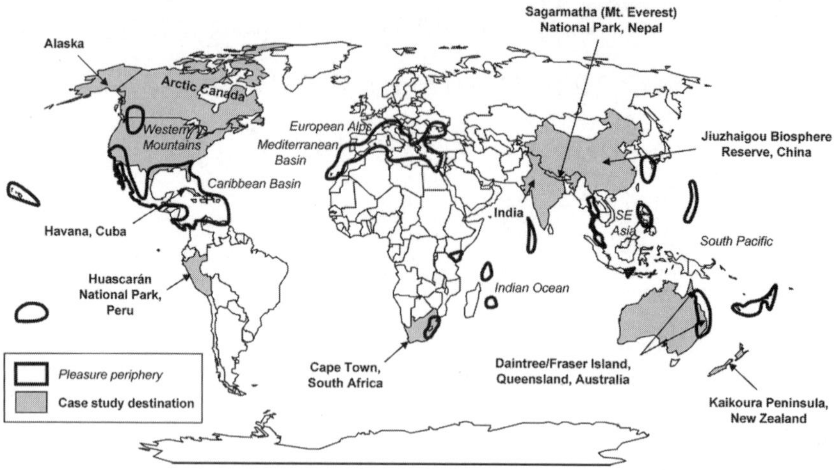

Figure 1.1 Location of the 'pleasure periphery', and of destinations and destination regions that feature as case studies in this book

Source: Adapted from Weaver (1998: 48)

exist (Orams 2001, Donohoe and Needham 2006) but, in its broadest sense, it concerns travel to a natural area; involving local people; feeding economic profit into local environmental protection; and contributing to the maintenance of the local environment and species diversity through minimizing visitor impact and promoting tourist education (Valentine 1993, Western 1993, Ceballos-Lascuráin 1998, Diamantis 1999, Fennell 2001, 2003). In this way, ecotourism (as a subset of alternative tourism) is being promoted by governments and the tourism industry as a sustainable alternative to mass tourism. However, critics have suggested that ecotourism can be damaging to the natural environment, not least with respect to the environmental cost of air travel to popular destinations and the potential for unwittingly disturbing soils and ecosystems through accommodation and activities on site (Wheeller 1991, 1993, Hjalager 1996, Conservation International 1999, Wearing and Neil 1999, Kruger 2005). Doubts have also been raised about the 'true' motivations of ecotourists, which appear to have as much to do with sustaining the ego as the environment (Wheeller 1993), and the sense in which ecotourism encourages increased use of natural areas and greater penetration into sensitive environments, thereby putting the very future of indigenous tourism industries at risk (Mihalic 2000). There is a tendency to overstate its significance, too, when in fact it constitutes only a small proportion of global tourism (estimated at between 2 per cent and 4 per cent; WTO 2002, cited in Cater 2004).

Before we explain the aim, philosophy and structure of this book, it is perhaps useful to provide a brief introduction to definitions of ecotourism; ecotourist types and motivations; and ecotourism's (environmental) sustainability credentials.

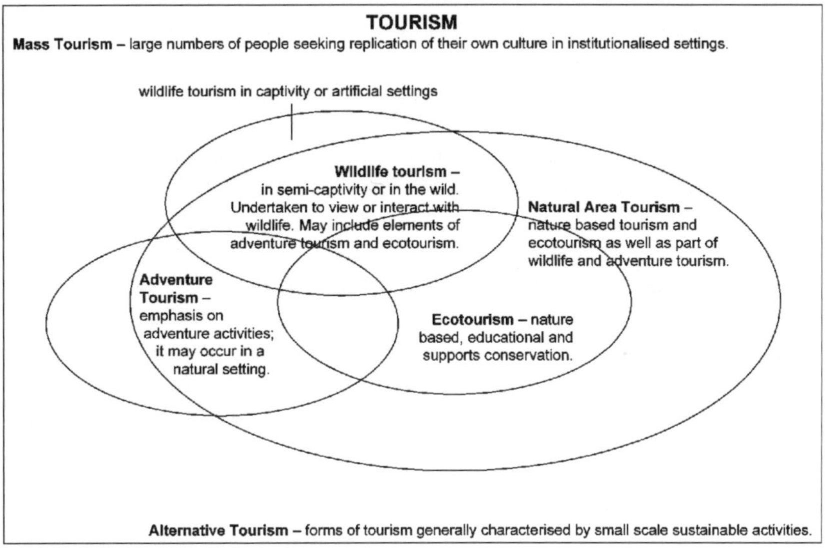

Figure 1.2 Relationship of ecotourism to other forms of tourism

With regards to the first consideration, Fennell's (2001) content analysis of ecotourism definitions identifies no less than 85 of these in the literature available at that time which, for a term that only entered academic discourse in the late 1980s, is quite remarkable (if not a little counter-productive). Definitions of ecotourism leave much to the interpretation of the reader, but they more or less cohere around three criteria, namely that '(1) attractions should be predominantly nature-based, (2) visitor interactions with those attractions should be focused on learning or education, and (3) experience and product management should follow principles and practices associated with ecological, socio-cultural and economic sustainability' (Weaver and Lawton 2007: 170). However, given the existence of something that approximates to an ecotourism experience in modified settings such as botanic gardens (see Chapter 12), not to mention the 'accidental ecotourist' and the potential for behaviour associated with states of mindlessness as well as mindfulness (after Moscardo 1996, 1999), plus an emerging critique of earlier work which problematizes ecotourism's role in sustainable development and environmental conservation (for example Butcher 2005, Kruger 2005, Cusack and Dixon 2006, Southgate 2006), one has to accept that there is no definition available that satisfactorily explains this phenomenon of contemporary (postmodern) society. That said, the distinction may usefully be made here between positive (i.e. what is) and normative (i.e. what should be) ecotourism definitions; whereas a universal definition that everyone can subscribe to remains a distant prospect (and, arguably, a futile project), there is an emerging consensus over the inclusion of certain 'value-based dimensions' such as conservation, community involvement and social responsibility (Weaver and Lawton, 2007: 1169). This is also apparent in Garrod's (2003) Delphi study, which indicates a preference amongst experts for prescriptive, rather than descriptive, definitions of ecotourism (as well as summarizing the reasons why they can be useful).

Figure 1.2 illustrates the relationship between ecotourism and other forms of tourism. This follows the typology of Newsome et al. (2002) in which ecotourism is seen entirely as a subset of natural area tourism, incorporating elements of wildlife and adventure tourism as acknowledged by a number of researchers (for example Page and Dowling 2002, Fennell 2003). It accommodates Fennell's (2003: 16) belief 'that ecotourism is distinct from mass tourism and various other forms of AT [alternative tourism]', whilst acknowledging the common ground between them (see Weaver 2005 on mass ecotourism, and Fennell 2003 on the difficulty of distinguishing between adventure tourism, cultural tourism and ecotourism in similar environmental settings). However, by its inclusion we are not asserting that this is the only way of seeing ecotourism in the context of other tourism types (see Weaver 2001 for alternative illustrations).

Moving on to the second of the above-mentioned considerations, namely ecotourists and their motivations, empirical research has shown that those who have participated in some form of nature-based tourism (which qualifies them as 'ecotourists' in a number of published studies) tend to be slightly older (between 35 and 54 in particular), better educated and more affluent (and, therefore, prepared to pay more for their holidays) than those who do not participate (see Wilson and Garrod 2003 for a summary of demographic and other characteristics). Furthermore, they are likely to stay longer and be more tolerant of basic conditions than general travellers. Environmental awareness and respect for local customs and culture are also differentiating features, as is gender (at least for certain activities such as bird watching, which attracts more males than females). However, despite having much in common, there are many different types of eco-/nature tourist that are distinguishable by factors including level of organisation (such as do-it-yourself ecotourists, ecotourists on tours and school or scientific groups, as identified by Kusler 1991) and dedication to nature (for example Lindberg's 1991 typology, comprising hard-core, dedicated, mainstream and casual nature tourists). Similarly, the degree of difficulty and level of engagement are used by Laarman and Durst (1987) in arguing for the distinction between 'hard' and 'soft' ecotourism experiences.

The most common reasons offered by ecotourists for undertaking a given trip are a desire to enjoy scenery and nature, and to encounter new places and experiences (see Chapters 9 and 11), with the imperative to 'get away from it all' of less importance to the ecotourist than, say, the adventure tourist. Motivations can and do vary by market, however, as summarized by Wight (2001). Suffice to say, for an *explanation* as to what motivates the ecotourist, one sometimes has to look beyond the academic literature. For example, in the first quarter of 2003 the British Broadcasting Corporation (BBC) conducted a survey of visitors to its website that attempted to establish the '50 things to do before you die', the

Table 1.1 '50 things to do before you die'

1	*Swim with dolphins*	11	Walk the Inca Trail	21	Travel on the Orient Express	31	Gallop a horse along a beach	41	Fish for Blue Marlin in the Florida Keys
2	*Dive the Great Barrier Reef*	12	Climb Sydney Harbour Bridge	22	*Bathe an elephant*	32	Explore the pyramids by camel	42	Go paragliding
3	Fly on Concorde	13	Escape to a paradise island	23	*Visit Antarctica*	33	Travel on the Trans-Siberian Railway	43	Play a round of golf at Augusta, Georgia
4	*Go whale-watching*	14	Drive a Formula 1 Car	24	Ride a Harley Davidson on the open road	34	Watch the sunset at Uluru (Ayers Rock)	44	*Trek to see a mountain gorilla*
5	*Dive with sharks*	15	Go white-water rafting	25	Have a go at cowboy ranching	35	Go wing-walking	45	*See tigers in the wild*
6	Skydiving	16	Walk the Great Wall of China	26	Climb Mount Everest	36	Climb Mount Kilimanjaro	46	Hurtle down the Cresta Run
7	Fly in a hot air balloon	17	Bungee jumping	27	Experience a waterfall	37	Fly over a volcano	47	Visit Walt Disney World, Florida
8	Fly in a fighter jet	18	Take a train journey across the Canadian Rockies	28	See the Earth from space	38	Drive a husky sled	48	Visit Las Vegas casinos
9	*Go on safari*	19	Drive along Route 66	29	*Explore the Galapagos Islands*	39	Hike up a glacier	49	*Watch Orang-Utans*
10	See the Northern Lights	20	Fly in a helicopter over the Grand Canyon	30	*Trek through a rainforest*	40	Ride a rollercoaster	50	*Go polar bear watching*

Source: BBC (2003)

Note: Experiences with a significant ecotourism component are highlighted in *italics*.

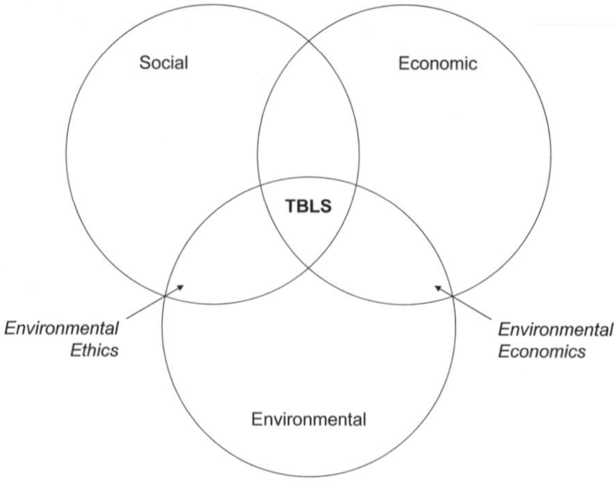

Figure 1.3 Elements of triple-bottom-line-sustainability (TBLS)

results of which were subsequently broadcast on the BBC's *Holiday* programme (Table 1.1). Not surprisingly, long-distance travel features in nearly all of these 'must-do' activities. What is more pertinent, however, is that a large number correlate with a wildlife or adventure tourism experience, most likely with a significant ecotourism component. Examples include, but are not confined to: swimming with dolphins (1); going on safari (9); seeing elephants in the wild (22); exploring the Galapagos Islands (29); trekking through a rainforest (30); and watching mountain gorillas (44). Crucially, these activities remind us of the potential for ecotourism (and related forms of tourism) to deliver what Maslow, writing in the 1950s, described as a 'peak experience'. In turn, they are able to address higher order human needs, especially self-actualization and transcendence (personal growth and fulfilment, and helping others to self-actualize). Indeed, opportunities to develop knowledge and understanding in beautiful surroundings (and, therefore, to satisfy cognitive and aesthetic needs as a pre-condition for self-actualization) are present in most, if not all, ecotourism encounters. Hence, we can conclude that ecotourists are motivated by so-called 'growth motivators' (and not just 'deficiency motivators', such as physiological, security, belongingness/love and esteem needs), and that ecotourism can help them to achieve their full potential (see Maslow 1987 for an explanation of these terms).

Finally, there is the issue of ecotourism and environmental sustainability (the latter being one of the components comprising the 'triple-bottom-line') (see Figure 1.3). Broadly speaking, ecotourism is linked to the emergence of 'new tourism' and associated production practices and consumption patterns (see Poon 1993); it is one of four key types identified by Shaw and Williams (2004: 119), the others being heritage/cultural tourism, adventure tourism, and visiting theme parks/mega-shopping malls (see also Mowforth and Munt 2003: 93 for an extended typology/glossary of terms). Although contested, it is suggested that some of these new forms of tourism are based around sustainable ideas (such as environmental stewardship, inter-generational and intra-generational equity). Certainly, we are told that ecotourism is small in scale, non-consumptive, ethical/responsible, and of benefit to local people. It is also thought to encourage pro-environmental behaviours (both home and away), when accompanied by interpretation (see Chapter 11).

This was the thinking behind the International Year of Ecotourism, in 2002 (as proclaimed by the United Nations). However, in the absence of an appropriate management regime, where 'those who initiate and develop ecotourism … operate within the capacity of the environment to absorb the impacts' (Wilson and Garrod 2003: 4), ecotourism is unlikely to be sustainable ecologically by any relevant measure. There is also the possibility that ecotourists (however well intentioned) might disturb the feeding and breeding patterns of wildlife, transmit diseases and modify habitats, just by being present in environmentally sensitive areas. In addition, a great many tourism products labelled with the prefix 'eco-' have few, if any, of the above characteristics (see Chapter 13), which suggests that 'ecotourism is being used [primarily] to meet economic objectives by promoting the quality of the environment to attract international tourists' (Holden 2005: 130), and not for the specific purpose of natural resource conservation (a minority of cases excepted). Finally, given that most ecotourism destinations are geographically remote in relation to the markets they serve, there is the issue of carbon expenditures associated with long-distance travel and their contribution to climate change (see Chapters 2 and 3). Clearly, then, ecotourism is not (and can never be) *synonymous* with environmental sustainability, contrary to what some advocates would have us believe.

The Ecotourism Body of Knowledge

Despite the title of this section, it is not our intention here to review comprehensively the literature on ecotourism, especially as some excellent reviews exist already (see Weaver and Lawton 2007, referenced throughout this opening chapter). Instead, we direct readers to the following:

- The ever-expanding range of introductory textbooks, which deal at length with the origins and concepts of ecotourism; where it lies within broader tourism types; its economic, environmental/ecological and socio-cultural impacts; the planning and management of ecotourism businesses; and ecotourism organizations and policies (see, for example, Wearing and Neil 1999, Page and Dowling 2002, Fennell 2003, Weaver 2008).
- More specialized publications that focus on certain aspects such as policy and planning (Fennell and Dowling 2003); underlying environmental ideologies linked to politics of north–south relations and social justice

(Duffy 2002, Butcher 2007); and economic and sustainable development in less developed countries (Weaver 1998, Honey 1999).
- The 500-plus articles of relevance in refereed journals, accessible via academic databases and search engines using appropriate keywords such as 'ecotourism', 'ecotourist' and so on (including those published in the dedicated *Journal of Ecotourism*, established in 2002).

In their review of the ecotourism literature, Weaver and Lawton (2007) identify three 'macro-themes' that may help contextualize the contents and approach of this book. Firstly, they note that the subject has been transformed over the past decade due to processes of de-differentiation (the blurring of the boundaries between ecotourism and other forms of tourism, as exemplified by upscale ecolodges and recreational hunting and fishing in protected areas) and segmentation (such as the emergence of whale watching and Antarctic tourism as sub-fields of enquiry). Secondly, they record ongoing efforts to make sense of ecotourism impacts, which follow three distinct research trajectories (four, in their original schema) addressing ecological, economic and ethical considerations (dubbed 'the 3Es' by Howie 2003). When arranged in this order, these form a continuum that at the one end is dominated by the natural sciences and their preferred methods of investigation, and at the other end the social sciences. Finally, they identify a profound 'North–South' (or, alternatively, 'West–Rest') division in both the sector and the literature. This is confirmed by the tendency for supply-side case studies (of venues, businesses, etc.) to be based in the developing world, in particular Central America, Sub-Saharan Africa and Southeast Asia (see Weaver 1998 for details), and for demand-side case studies (of markets, products, etc.) to be based in the developed world, from which much knowledge of ecotourism originates (the implications of which are explored in Cater 2006).

Organisation of the Book

With the contextual discussion complete, we now turn to the remainder of this book, which aims to offer conceptual and practical insights into the complex interactions between ecotourism and heterogeneous and dynamic natural environments (building on the foundation provided by the sum of the contributions to Buckley 2004). Although privileging environmental relationships (encompassing physical environment, flora and fauna), consideration is given to government policy, marketing by suppliers, consumer behaviour and visitor/environmental management as factors that mediate tourist interactions with the natural environment. Some of these factors correlate with 'strategically important' areas of ecotourism that have been neglected thus far in the literature (see Weaver and Lawton 2007).

The book is divided into three parts. The first part, comprising Chapters 1 to 3, introduces ecotourism and tourism–environment relationships at the local and global scales. The second part, comprising Chapters 4 to 12, is devoted to a diverse series of case studies from around the world (see Figure 1.1), incorporating analysis from a range of geographic scales and integrating natural– and social–scientific knowledge, yet resisting a noticeable trend to showcase best practice (given that this is rarely transferable from one context to another). A variety of ecotourism activities, ecosystem types, ecosystem components and environmental responses are examined in these chapters. The final part, comprising Chapters 13 and 14, draws together some of the lessons revealed by earlier chapters in a critical examination of eco-/nature tourism and its articulation with discourses of environmental sustainability, and includes an 'agenda' for future research in this area.

In Chapter 2, Andrew Holden considers the negative and positive impacts of tourism in general (not just ecotourism) on the natural environment, emphasizing the significance of human agency and of the values we attach to nature (which vary by cultural group and over time) in determining the net environmental impact of tourism. The roles of environmental economics and environmental ethics in promoting natural resource conservation are also discussed in this chapter. It provides a basis for subsequent chapters that explore the relationship of tourism with natural resources in theory and offer practical examples of resource conserving behaviour vis-à-vis resource destructive behaviour in eco- and related forms of tourism. Then in Chapter 3, Colin Hunter uses Ecological Footprint (EF) analysis, which has the ability to compare impacts generated in transit with those at the destination, to assess the environmental sustainability of ecotourism relative to mass tourism. He makes the critical distinction between gross and net EF estimates (with the latter taking into account the footprint that would have been generated had the tourist stayed at home), in outlining the circumstances in which return trips from some of the world's leading source countries/cities to recognized ecotourism destinations could be said to compare favourably with equivalent trips to mass market destinations (despite concerns about ecotourism's frequent reliance upon aviation).

The next nine chapters provide case studies of particular sites and destinations. In Chapter 4, Alton Byers undertakes a comparative analysis of landscape change in the alpine zones of Sagarmatha (Mt. Everest) National Park, Khumbu, Nepal and the Huascarán National Park, Cordillera Blanca, Peru. He attributes the degradation of the alpine zone within each national park to poorly controlled adventure tourism and associated impacts, and reports on progress made with regards to conserving, protecting and restoring these fragile environments. Tourism in protected areas also features in Chapter 5, by Wenjun Li. Using a case study of Jiuzhaigou Biosphere Reserve, located within China's Sichuan Province, she underlines the importance of community involvement in tourism and in ensuring that residents benefit financially from it, here by means of a 'joint-stock mechanism'. With such a mechanism in place, the likelihood of balancing resource conservation and economic development objectives through tourism is increased. Then in Chapter 6, Emma Stewart and Dianne Draper investigate the development of cruise tourism in Arctic Canada and the environmental consequences thereof. They identify a package of measures aimed at ensuring a successful and sustainable future for the Canadian Arctic as a cruise tourism destination, from the perspective of the stakeholders involved (communities, operators and tourists), and they discuss the likely implications of climate change for the industry in this part of the world.

The focus switches to wildlife tourism and particular species in the four chapters that follow. In Chapter 7, Terry DeBruyn and Tom Smith explore some of the issues raised by the growing demand for brown bear-viewing on Alaska's public lands, together with resource management interventions designed to minimize human impacts on bears and associated wilderness. This culminates in a series of guidelines for managing bear-viewing areas that prioritizes bear conservation, human safety and visitor satisfaction. Then, in Chapter 8, Simon Evans critically examines the role of ecotourism in protecting the wild tiger populations of India. He voices concern over the widespread practice of encouraging tigers to visit a site habitually through baiting, with a view to guaranteeing sightings. This situation is compounded by marketing approaches that heighten tourists' expectations of viewing and photographing tigers at close range (see Dobson's 2007 study of sharks for some interesting parallels) and, more seriously, declining tiger numbers as a consequence of large-scale poaching (which reduces the likelihood of a chance encounter). In contrast to the experience outlined in Chapter 5 (see above), he notes a tendency for local populations to be excluded from ecotourism and conservation activities, leading to the formation of eco-enclaves (the tiger reserves and the hotels located outside reserve boundaries) which capture most of the financial benefits of tiger-related tourism. Following this, in Chapter 9 Leah Burns draws on the testimony of visitors and park rangers, in a study of tourism impacts and tourist aspirations in relation to the dingoes of Fraser Island, Queensland, Australia. She evaluates hard (e.g. fences) and soft (e.g. visitor education) approaches to managing human–dingo interactions, in the wake of highly publicized incidents that resulted in people being harmed and dingoes being culled (and, like Simon Evans before her, draws attention to the problem of habituation in wildlife tourism).

The role and potential of interpretation provision in managing visitor behaviour and attitudes towards wildlife and ecosystem conservation is examined in the next two chapters. In Chapter 10, Davina Stanford uses a large-sample survey of visitors (n=372) to the fur seal colony at Kaikoura Peninsula, New Zealand to evaluate the effectiveness of six different messages designed to encourage the visitor not to get too close to the seals (less than ten metres), which correspond to Kohlberg's stages of moral development. The results indicate that stage four and five messages (emphasizing good citizenship, and the impacts and consequences of less than desirable behaviour) are the most influential. Then in Chapter 11, Jennifer Hill and Georgie Gough report the findings of a visitor survey conducted at Crocodylus Rainforest Village in Daintree, Australia (n=267). This provides insights into interpretation provision as a tool for educating the travelling public about biodiversity, including pertinent threats and conservation issues, and suggests that it engenders (or at the very least consolidates) positive attitudes and behavioural intentions towards ecosystem conservation specifically and to the environment more generally, as well as enhancing visitor satisfaction.

If the last two chapters (excluding the conclusion) have something in common, it is that they deconstruct the term 'ecotourism', firstly by extending it to encompass modified spaces and, secondly, by dismissing it as a mere marketing label (so-called 'greenwashing'). In Chapter 12, Lucy A. Sutherland examines the potential role of botanic gardens in less developed countries (here represented by South Africa and Cuba) in linking into important urban green space and wildlife corridor networks, and in alleviating pressure on adjacent protected areas by providing a substitute ecotourism experience. Using a combination of questionnaires distributed to and semi-structured interviews with botanic garden visitors and inbound tour operators, she identifies the factors that, in combination, will determine whether or not these attractions can be incorporated within ecotourism programmes (as a more sustainable alternative to visiting fragile natural environments). This is followed by Ken Simpson's polemical essay, in Chapter 13, on the theoretical foundations of ecotourism and the political, societal and ecological realities that militate against them. He challenges the notion that 'if it's ecotourism, it must be sustainable' by arguing that the cost-benefits associated with any given form of tourism are the product of complex interactions between a destination with experiences to sell and a multi-faceted market with the resources to buy, rather than being fixed or given. Indeed, it would seem that 'ecotourism' is in danger of becoming little more than a brand, due to its appropriation by tourism operators and destination marketers. To conclude, the key issues raised by the book are summarized in Chapter 14, and a number of directions for ecotourism (and environmental sustainability) research are proposed, informed by wider developments in tourism studies and the social sciences.

References

Ayala, H. (1996) Resort ecotourism: a paradigm for the 21st century. *Cornell Hotel and Restaurant Administration Quarterly*, 37, 46–53.

British Broadcasting Corporation (BBC) (2003) *50 Things To Do Before You Die*. Available at: http://www.bbc.co.uk/50/destinations/things [accessed: 19 September 2003].

Boo, E. (1990) *Ecotourism: The Potentials and the Pitfalls* (Vol. 1). World Wildlife Fund, Washington, DC.

Buckley, R. (2000) Tourism in the most fragile environments. *Tourism Recreation Research*, 25, 31–40.

Buckley, R. (2003) *Case Studies in Ecotourism*. CABI Publishing, Wallingford.

Buckley, R. (ed.) (2004) *Environmental Impacts of Ecotourism*. CABI Publishing, Wallingford.

Butcher, J. (2005) The moral authority of ecotourism: a critique. *Current Issues in Tourism*, 8, 114–124.

Butcher, J. (2007) *Ecotourism, NGOs and Development: A Critical Analysis*. Routledge, Abingdon.

Cater, E. (2004) Ecotourism: theory and practice. In A.A. Lew, C.M. Hall and A.M. Williams (eds) *A Companion to Tourism*. Blackwell Publishing, Oxford, pp. 484–497.

Cater, E. (2006) Ecotourism as a Western construct. *Journal of Ecotourism*, 5, 23–39.

Cater, E. and Lowman, G. (eds) (1994) *Ecotourism: A Sustainable Option?* John Wiley & Sons, Chichester.

Ceballos-Lascuráin, H. (1998) Introduction. In K. Lindberg, M. Epler-Wood and D. Engeldrum (eds) *Ecotourism – A Guide for Planners and Managers* (Vol. 2). The Ecotourism Society, North Bennington, Vermont, pp. 7–10.

Conservation International (1999) *The Green Host Effect – An Integrated Approach to Sustainable Tourism and Resort Development*. Conservation International, Washington, DC.

Cusack, D. and Dixon, L. (2006) Community-based ecotourism and sustainability: cases in Bocas del Toro Province, Panama and Talamanca, Costa Rica. *Journal of Sustainable Forestry*, 22, 157–182.

Diamantis, D. (1999) The concept of ecotourism: evolution and trends. *Current Issues in Tourism*, 2, 93–122.

Dobson, J. (2007) Shark! A new frontier in tourist demand for marine wildlife. In J. Higham and M. Lück (eds) *Marine Wildlife and Tourism Management: Insights from the Natural and Social Sciences*. CABI Publishing, Wallingford, pp. 49–65.

Donohoe H.M. and Needham, R.D. (2006) Ecotourism: the evolving contemporary definition. *Journal of Ecotourism*, 5, 192–210.

Duffy, R. (2002) *A Trip Too Far: Ecotourism, Politics and Exploitation*. Earthscan, London.

Fennell, D. (2001) A content analysis of ecotourism definitions. *Current Issues in Tourism*, 4, 403–421.

Fennell, D. (2003) *Ecotourism: An Introduction* (2nd edition). Routledge, London.

Fennell, D. and Dowling, R.K. (eds) (2003) *Ecotourism Policy and Planning*. CABI Publishing, Wallingford.

Garrod, B. (2003) Defining marine ecotourism: a Delphi study. In B. Garrod and J.C. Wilson (eds) *Marine Ecotourism: Issues and Experiences*. Channel View Publications, Clevedon, pp. 17–36.

Hjalager, A. (1996) Tourism and the environment: the innovation connection. *Journal of Sustainable Tourism*, 4, 201–218.

Holden, A. (2005) *Tourism Studies and the Social Sciences*. Routledge, London.

Honey, M. (1999) *Ecotourism and Sustainable Development: Who owns Paradise?* Island Press, Washington, DC.

Howie, F. (2003) *Managing the Tourist Destination*. Continuum, London.

Kruger, O. (2005) The role of ecotourism in conservation: panacea or Pandora's box? *Biodiversity and Conservation*, 14, 579–600.

Kuo, I-L. (2002) The effectiveness of environmental interpretation at resource-sensitive tourism destinations. *International Journal of Tourism Research*, 4, 87–101.

Kusler, J.A. (1991) Ecotourism and resource conservation: introduction to issues. In J.A. Kusler (ed.) *Ecotourism and Resource Conservation: A Collection of Papers* (Vol. 1). Omnipress, Madison, WI.

Laarman, J. and Durst, P. (1987) *Nature Travel and Tropical Forests. FREI Working Paper Series*. Southeastern Center for Forest Economics Research, North Carolina State University, Raleigh, NC.

Lindberg, K. (1991) *Policies for Maximising Nature Tourism's Ecological and Economic Benefits*, World Resources Institute, Washington, DC.

Maslow, A.H. (1987) *Motivation and Personality* (3rd edition). Harper & Row, New York, NY.

Mihalic, T. (2000) Environmental management of a tourist destination: a factor of tourism competitiveness. *Tourism Management*, 21, 65–78.

Moscardo, G. (1996) Mindful visitors: heritage and tourism. *Annals of Tourism Research*, 23, 376–397.

Moscardo, G. (1999) *Making Visitors Mindful*. Sagamore Publishing, Champaign, IL.

Mowforth, M. and Munt, I. (2003) *Tourism and Sustainability: New Tourism in the Third World* (2nd edition). Routledge, London.

Newsome, D., Moore, S.A. and Dowling, R.K. (2002) *Natural Area Tourism: Ecology, Impacts and Management*. Channel View Publications, Clevedon.

Newsome, D., Dowling, R.K. and Moore, S. (2005) *Wildlife Tourism*. Channel View Publications, Clevedon.

Orams, M.B. (2001) Types of Ecotourism. In D.B. Weaver (ed.) *The Encyclopaedia of Ecotourism*. CABI Publishing, Wallingford, pp. 23–36.

Page, S.J. and Dowling, R.K. (2002) *Ecotourism*. Prentice Hall, Harlow.

Poon, A. (1993) *Tourism, Technology and Competitive Strategies*. CABI Publishing, Wallingford.

Ryan, C., Hughes, K. and Chirgwin, S. (2000) The gaze, spectacle and ecotourism. *Annals of Tourism Research*, 27, 148–163.

Southgate, C.R.J. (2006) Ecotourism in Kenya: the vulnerability of communities. *Journal of Ecotourism*, 5, 80–96.

Shaw, G. and Williams, A.M. (2004) *Tourism and Tourism Spaces*. Sage, London.

The International Ecotourism Society (TIES) (2006) *Fact Sheet: Global Ecotourism*. Available at: http://www.box.net/shared/static/eaukonjc0a.pdf [accessed: 13 August 2008].

Turner, L. and Ash, J. (1975) *The Golden Hordes: International Tourism and the Pleasure Periphery*. Constable, London.

Valentine, P.S. (1992) Review. Nature-based tourism. In B. Weiler and C.M. Hall (eds) *Special Interest Tourism*. Belhaven Press, London, pp. 105–128.

Valentine, P.S. (1993) Ecotourism and nature conservation: a definition with some recent developments in Micronesia. *Tourism Management*, 14, 107–115.

Wearing, S. and Neil, J. (1999) *Ecotourism: Impacts, Potential and Possibilities*. Butterworth Heinemann, Oxford and Melbourne.

Weaver, D.B. (1998) *Ecotourism in the Less Developed World*. CABI Publishing, Wallingford.

Weaver, D.B. (2001) Ecotourism in the context of other tourism types. In D.B. Weaver (ed.) *The Encyclopaedia of Ecotourism*. CABI Publishing, Wallingford, pp. 73–83.

Weaver, D.B. (2005) Mass and urban ecotourism: new manifestations of an old concept. *Tourism Recreation Research*, 30, 19–26.

Weaver, D. (2008) *Ecotourism* (2nd edition). John Wiley & Sons, Chichester.

Weaver, D.B. and Lawton, L.J. (2007) Twenty years on: the state of contemporary ecotourism research. *Tourism Management*, 28, 1168–1179.

Western, D. (1993) Defining ecotourism. In K. Lindberg and D.E. Hawkins (eds) *Ecotourism: A Guide for Planners and Managers*. The Ecotourism Society, North Bennington, Vermont, pp. 7–11.

Wheeller, B. (1991) Tourism's troubled times: responsible tourism is not the answer. *Tourism Management*, 12, 91–96.

Wheeller, B. (1993) Sustaining the ego. *Journal of Sustainable Tourism*. 1, 121–129.

Wight, P. (2001) Ecotourists: not a homogenous market segment. In D.B. Weaver (ed.) *The Encyclopaedia of Ecotourism*. CABI Publishing, Wallingford, pp. 37–62.

Wilson, J.C. and Garrod, B. (2003) Introduction. In B. Garrod and J.C. Wilson (eds) *Marine Ecotourism: Issues and Experiences*. Channel View Publications, Clevedon, pp. 1–11.

Chapter 2
An Introduction to Tourism–Environment Relationships

Andrew Holden

Introduction

In attempting to understand the relationship of tourism with the natural environment it is necessary to reflect that the tourism system incorporates a range of stakeholders, including governments, the private sector, non-governmental organisations (NGOs), communities and tourists. The common denominator of these stakeholders is that they all involve human agency and behaviour. Consequently, when analysing the relationship between tourism and the natural environment it is necessary to consider that the relationship is in fact an expression of human interaction with nature. Subsequently, the values that we attach to nature, our cultural construction of what it is and our ethical relationship with it, will be influential for determining the impacts of tourism upon it. This chapter explores these issues, explaining how tourism has negative and positive consequences for the natural environment and, through a reciprocal relationship, how tourism itself can be threatened by human-induced changes in the environment. The chapter then progresses to an analysis of the ethical relationship between tourism and the natural environment.

Impacts of Tourism on the Natural Environment

To provide context for exploring the impacts of tourism on nature, it is useful to think of tourism as a system that links the environments of places that people come from, with the ones that they go to. Although the focus of this chapter is not upon the influence that the home environment has on demand for tourism, it is evident that many tourists seek what are perceived to be 'unspoilt' or 'natural' environments in contrast to the urban environments they often come from. The mental construction of the 'natural' environments we desire to travel to is an essential component of tourism. The home and destination environments are linked by the transport component of tourism, of which there is an increasing focus upon its environmental impacts, notably the contribution of aviation to global warming. Within this system, the environmental impacts of tourism may subsequently be viewed as having varying spatial dimensions, from the local to the global. The impacts may also be differentiated into two broad categories of 'negative' and 'positive'; that is determined as being harmful or beneficial to nature.

The evaluation of the balance of this relationship, between the positive and the negative, is reflective of the values attributed to nature. If nature is categorized as possessing primarily an instrumental value for the purpose of human benefit, then there is a propensity for its over-use. This likelihood is enhanced by the characteristics of some of the natural resources that tourism relies upon, which are representative of 'common pool resources' (CPRs). According to Ostrom et al. (1999), the term CPR refers to resources with various types of property rights, but where exclusion of beneficiaries is impractical on the basis of cost, or is at least very costly, and the exploitation of the resource by one person reduces the benefit for another.

A subsequent major threat to the well-being of CPRs exists from the mentality of 'finders-keepers', that is a rush to harvest and secure the benefits of the resource before someone else does (Hardin 1968). Typical CPRs used for tourism may include the oceans, the atmosphere, beaches, coral reefs, wetlands, forests, wildlife and mountains. CPRs are vulnerable to over-use (what is taken from them) and also to pollution (what is put into them).

Conversely, if nature is considered to possess a range of values beyond the instrumental, then our interaction with it may be more benign. Certainly, if the conservation of nature through tourism can be demonstrated to have an economic value, then from a development perspective the argument for its continued existence is stronger. The next section of the chapter considers the negative and positive environmental impacts of tourism.

Negative Impacts: Environmental Pollution, Ecosystem Degradation and Reduced Environmental Aesthetic

As stated in the introduction, the spatial dimension of the environmental impacts of tourism range from the global to the local. At a global spatial scale, the contribution of air transport to pollution and global warming is an issue of debate and media attention. Although the amount of CO_2 emissions from aviation is relatively low compared to other sources: the International Air Transport Association (IATA 2007) suggest it is 2 per cent, whilst Gössling and Peeters (2007) estimate it to be between 3.4 per cent to 6.8 per cent; it is the rapid growth in air transport and its increasing significance as a source of CO_2 emissions that is of concern. Indeed, several climatologists state that aviation is the fastest growing causal factor of climate change (Garman 2006) and, per passenger kilometre, aviation produces more CO_2 than any other form of transport (Dubois and Ceron 2006). The potential impacts of greenhouse gas emissions from aircraft are made more significant by the effects of 'radiative forcing', i.e. as they are emitted at altitude, aircraft emissions have a greater relative impact than they would at surface level. The IPCC (2007) estimates the net effect of these emissions, taking into account radiative forcing, to be 2.7 times the carbon consumed in fuel.

Concern over air transport is a consequence of its exponential growth since the 1950s, with the annual rate of passenger growth averaging 5 per cent to 6 per cent per annum over almost a 50-year period. This demand has more recently been 'fuelled' by a combination of rapid growth in the numbers of budget flights and rising demand for flights in countries with rapidly developing economies. For example, in the case of China, during the period of May 2006 to May 2007, domestic flights increased by 18 per cent, whilst international flights to and from the country increased by 17 per cent (McCarthy 2007). The extent to which technology can provide a solution to increasing CO_2 emissions is uncertain, as the exponential increase in demand for flying means that total aviation emissions continue to rise despite improvements in emission-reducing technology (IPCC 2007, Adams 2007). In the absence of major-step technological progress, such as hydrogen or biofuel powered commercially viable aircraft, aviation emissions will likely be a major contributor to greenhouse gas emissions in the future. Even if alternative fuel sources were available, their medium- to long-term environmental impacts are uncertain, and the removal of land from agricultural use for biofuel production is already having adverse economic effects leading to rising food prices and shortages.

Whilst aviation, and other carbon-based modes of transport used for tourism and tourism infrastructure, generally makes a contribution to global warming, in a reciprocal relationship, global warming poses a threat to the future of tourism. The World Tourism Organisation (2003: 8) state: 'In two environments that are vital for tourism activities and where tourism is an equally vital component in regional and local economies – coastal zones and mountain regions – climate change puts tourism at risk'. Thus, environmental change can threaten the very nature of the destinations that tourists seek to visit using air travel (Holden 2005).

Alongside its global effects, tourism impacts upon the natural environment at a local level. Tourism development can directly threaten and destroy ecological habitats, although the scale of transformation is difficult to estimate in the absence of comprehensive detailed research. There is, however, empirical evidence that tourism can impact upon the functioning of ecosystems, for example coral reefs, which are broadly comparable in terms of species diversity to tropical rain forests (Hill and Hill 2001). Coral is threatened from its use for the construction of tourism facilities, inadequate sewage disposal from proximate tourism concerns and the operations of local tour enterprises. For example, coral is mined for building materials in Sri Lanka, India, the Maldives, east Africa, Tonga and Samoa (Mieczkowski 1995). In the Red Sea, sediment run-off as a by-product of tourism construction, combined with nourishment to enhance tourist beaches and poor quality sewage and rubbish disposal systems for tourism, threaten the well-being of an estimated 70 per cent of the reefs off the coast of Egypt (Goudie and Viles 1997). In turn, these changes challenge the viability of the area's lucrative diving industry, which is dependent upon the attractive biodiversity of the reefs.

Further problems associated with the use of coral reefs are linked to tourist behaviour, including tourists walking upon it, divers breaking it through accidental contact or purposefully for souvenirs, tour operators dragging their boat anchors through it and local people breaking it to sell or make handicrafts with. These activities result in mechanical damage to reefs, altering coral cover and morphology, and impacting upon coral diversity and productivity (Palaganas 1991, Hawkins and Roberts 1993, Rouphael and Inglis 1997, Jameson et al. 2007). Although tourism may pose a notable threat to coral reefs in some areas, it is important to keep its impacts in perspective. At a global level, other factors, including global warming, tropical storms, increased sedimentation from deforestation, industrial pollution and over-fishing, also pose significant threats to reefs (Holden 2005).

Another example of the pressure that tourism can place on ecosystems includes the deforestation of mountainsides for ski development. Mountain areas harbour ecosystems that are sensitive to change, with forest removal not only potentially resulting in a loss of habitat for wildlife, but also leading to increased overland flow after rain and reduced soil cohesion. The combination of increased overland flow and reduced soil cohesion can heighten the propensity for landslides and in some cases lead to subsequent loss of life. The deaths, for example, of 60 people in northern Italy and southern Switzerland in 1987 were caused by landslides, which were attributed to deforestation for ski development (Simons 1988).

Coastal wetlands, an ecosystem that covers 6 per cent of the earth's surface and is significant for its biodiversity, carbon storing qualities and flood control capabilities, may also be placed under pressure from tourism development (Holden 2008). The restricted availability of land for tourism development in coastal and island areas, combined with the engineering capability to reclaim wetlands, means that they are increasingly being used for tourism development. For example, the Sir Donald Sangster International Airport in Jamaica and the French Mediterranean resort of 'La Grande Motte' were built on reclaimed wetlands (Holden 2008). The loss of wetland ecosystems not only leads to a loss of flora and fauna, but also decreases the ability of land areas to cope with flooding and reduces the capacity of the earth's 'carbon sink'.

Alongside the problem of aircraft emissions already alluded to, other types of pollution related to tourism include water, noise and a reduced environmental aesthetic. Water pollution is a major problem in many tourist regions of the world. For instance, in the most visited tourist area of the world, the Mediterranean, only 30 per cent of over 700 towns and cities on the coastline treat sewage before discharging it into the sea (Jenner and Smith 1992). In the Caribbean Basin, where tens of millions of tourists annually join the 170 million inhabitants, only 10 per cent of the sewage is treated before being discharged into the sea. Although these figures for sewage treatment may appear low, compared to other areas of the world they are actually good. Other regular international tourist destinations such as east Asia, east Africa and the islands of the South Pacific, with a few exceptions, have either no sewage treatment or treatment plants that are totally inadequate for the size of the population (Jenner and Smith 1992).

Although nearly two decades after the writing of Jenner and Smith (1992) it could be expected that these figures should have improved, it is suggested that the extent of any improvement is likely to be irregular, dependent upon political will and economic resources to mitigate the effects of pollution. The problem of water contamination from human sewage is not caused exclusively by tourism, but is reflective of an inadequate infrastructure to meet the needs of both local populations and tourists.

Alongside the consequences it can have for human health, causing diseases ranging from mild stomach upsets to typhoid through the intake of water contaminated by faeces, human sewage also causes eutrophication (nutrient enrichment) of water. The consequence of a serious health scare for tourism demand was dramatically demonstrated in 1988 in Spain, when the fear of a typhoid outbreak resulting from contaminated water led to a 70 per cent decline in tourist reservations the next year (Kirkby 1996). The process of eutrophication may also impact on tourism demand, as was the case on the Romagna coast of Italy, where the total number of tourist bookings received by accommodation establishments in the area fell by 25 per cent in 1989 compared to 1988, owing to the eutrophication of the Adriatic and the spread of algae on the surface of the water (Becheri 1991).

In addition to its contribution to air pollution, a further negative externality of air traffic is noise pollution (Mieczkowski 1995). Noise pollution is a particular problem for those residents who are juxtaposed to busy international and domestic airports, and the proposed development of airports may sometimes lead to opposition by local people and environmental protest groups, for example Narita Airport in Tokyo in the 1990s (Shaw 1993) and, more recently, Heathrow in London. Noise pollution from tourism may be particularly noticeable in destinations where tourists are searching for quietness and peace. Air flights in remote areas where a lack of noise is expected, such as the Grand Canyon in the US and the Himalayas, can cause disruption to tourists and recreationists. Noise pollution from the construction of tourism facilities can also be a problem for residents and tourists. Briguglio and Briguglio (1996) observe that intense noise is generated by the building of hotels and other construction activity in destinations. Moreover, night clubs open until the early morning and increased car traffic from tourism movements add to the noise pollution experienced by both residents and tourists in tourism destinations.

Although possibly more subjective than necessarily scientific, a further category of pollution is ‘aesthetic pollution’. Commenting on the development of tourism in the Guadeloupe and Martinique islands in the Lesser Antilles, Burac (1996: 71) observes: ‘The most worrying problem now prevalent in the islands relates to the anarchic urbanisation of the coasts ... Also, the built-up areas by the seaside are often not aesthetically attractive due to the diversity of architectural styles, the disappearance of traditional creole homes and the disorderly way in which public posters are displayed’. Similarly, four decades ago Mishan (1969: 141) wrote: ‘Once serene and lovely towns such as Andorra and Biarritz are smothered with new hotels and the dust and roar of motorised traffic’.

Whilst it would be unrealistic to assume that tourism may develop without involving the transformation of nature, it is misleading to portray tourism as having a purely negative effect upon the natural environment. Similar to most other types of economic activity, tourism involves some degree of trade-off between the realization of human benefits and environmental costs. The demand for tourism can play a vital role in giving an economic value to a natural resource to remain in a conserved state. Tourism may also play a significant role in raising awareness to support the establishment of protected areas to conserve nature. This chapter now turns its attention to exemplifying such issues.

Positive Impacts: The Case for Tourism as an Agent of Conservation

Whilst nature has been a key component of recreational tourism since the epoch of Romanticism, the emergence of recent terminology, including ecotourism, nature tourism, sustainable and responsible tourism underlines the central role of nature to tourism. Although controversy exists over what these terms actually mean and how they should be practised, they are nevertheless significant to the context of thinking of tourism as an agent for conservation.

In contrast to other kinds of economic land-use, such as agriculture, forestry and mining, which typically require the modification or destruction of natural resources to realize their utility, user-benefits may be achieved through tourism from experiencing nature in a largely conserved state. The fact that tourists are willing to pay to travel to, and to use accommodation and other services to view what is held to be 'nature', brings varying degrees of economic benefits at a national and local level to a destination. Consequently, an economic value may be given to natural resources to remain in a conserved state, the link to sustainable tourism development being axiomatic, as suggested more generally by Pearce (1993: 15):

> If, on the other hand, conservation and the sustainable use of resources can be shown to be of economic value, then the dialogue of developer and conservationist may be viewed differently, not as one of necessary opposites, but of potential complements.

Whilst comparative estimates of the economic value of natural resources in a conserved state that can be realized through tourism are limited, research in the Amboseli National Park in Kenya in the late 1980s found the economic value of a lion to be US$27,000 per annum, measured against the tourists' willingness to pay to see it in its natural habitat (Boo 1990). In terms of relative economic use, the park's net earnings from tourism were found to be 50 times higher than the most optimistic projection for agricultural use.

The potential economic value of ecotourism vis-à-vis other development options was also emphasized in the case of the Mabira Forest Reserve in Uganda. Recognized for its biodiversity, the forest was under threat of removal for sugar cane plantation to produce ethanol. However, an economic analysis of the value of ecotourism and carbon capture calculated an annual value of US$316 million compared to US$20 million for ethanol production (Smith 2007).

Although the revenues that can be obtained through tourism can help to protect the environment from other more environmentally destructive development alternatives, an important caveat of this argument is that if natural habitats or wildlife are judged not to have sufficient economic value in comparison to other development options, then a pretext is established for their removal.

Shackley (1996) highlights a further danger of the economic valuation process for the conservation of wildlife. There are many species of wildlife which are not attractive to tourists in the dramatic sense that elephants and lions may be, but they have an essential role to play in the functioning of the ecological system of the area. Invertebrates, for example, perform essential ecosystem functions, yet they are largely overlooked in terms of ecotourism planning (Huntly et al. 2005). Because such species are not on the tourists' itinerary of animals to view, their value will be undermined in economic terms, placing their continued survival under threat.

Where tourism takes place in spatially defined boundaries, such as national parks or other types of protected areas that possess entry points and an evident management structure, the application of economic methodology to assess the value of wildlife that can be realized through tourism will be easier to implement than in areas which are less well defined. The use that revenues from tourism can have in protected areas is varied, including the conservation of natural resources, funding for infrastructure, employment opportunities, supplementary livelihood opportunities and human development opportunities. In the Annapurna Conservation Area of Nepal, for example, tourism revenues have enhanced economic development, whilst also being used to aid conservation programmes (Holden and Ewen 2004).

The vulnerability of tourism demand to external forces, however, means that basing protected area policy upon revenues from tourism is an unstable and potentially risky option. External factors that influence the demand for tourism to protected areas include security concerns, competition from other national parks and destination areas, and changing market fashions (Font et al. 2005).

Alongside methodology aimed at calculating the value of the environment, policy instruments can also be adopted to help give recognition to the value of conserving environments. One way of aiding the conservation of natural resources in developing countries is through 'debt-for-nature swaps'. The basis of this approach is that a conservation-based, non-governmental organisation (NGO) buys off an amount of a country's national debt, in return for guarantees that the indebted country will support and manage the conservation of a designated area, such as a national park. This will usually involve the development of appropriate management plans for the park. In effect, the NGO is purchasing the 'development rights' or, alternatively, the 'no-use' rights to a particular area of a country. As Pearce (1993) points out, debt-for-nature swaps are the only way in which estimates of the 'existence' (that is non-use) value of nature have been established.

According to Stone (1993), the concept of the debt-for-nature swap is an extension of the proposal made by Jomo Kenyatta, the founder of modern Kenya, at a conference in 1961. He intimated at the time that if African wildlife was a world possession, then 'the world could pay for it'. The process is based upon the buying of national debts on the world's money market at a fraction of their face value; for example, an NGO may pay a bank 25 cents for every dollar of a government debt. Banks are willing to discount the value of the loan and debt because of the risk that an indebted country may default upon its payment completely. As a consequence, debt-for-nature swaps usually involve the co-operation of national government, international and local NGOs and banks. In essence, a government is giving up part-sovereignty of a designated area of its country to an NGO.

Environmental Ethics

Debt-for-nature swaps are reflective of a philosophy that recognizes values in nature that transcend purely the instrumental and economic. A theme of this chapter is that as tourism is a phenomenon of human agency, how the tourism–environment relationship manifests itself will be reflective of how stakeholders value nature. The values that are held about nature extend beyond the economic, to incorporate the recreational and aesthetic; values that are especially significant in the case of tourism. Holmes Rolston III (1988) recognizes a range of other values that are held by nature, including the scientific, genetic, historical, cultural–symbolic and character-building. At a time when changes in the surrounding environment are forcing us to reconsider our position relative to nature, tourism as a sub-system of society will also adjust correspondingly to reflect new environmental considerations, priorities and ethics.

The relevance of environmental ethics to the relationship between tourism and nature can be illustrated through the following question: 'Should safaris be based upon the "shooting" of wildlife, as they predominantly were at the beginning of the twentieth century, or the "observing" of wildlife, as they are now?' In attempting to answer this question it is necessary to consider our relative position to nature and our consequential 'duty' towards it.

Reflections on the human position relative to nature were expressed by the influential western philosopher René Descartes in the seventeenth century. Arguing that unlike humans, animals possessed no souls and minds, he believed animals to be irrational and insensible making them incapable of suffering. This separation of the human and non-human worlds is dualistic, creating a pretext for the use of nature to satisfy human desires and aspirations. Descartes encouraged the belief that humans are the masters of nature. The dualism inherent to Cartesian philosophy was also advocated by another seminal thinker of western thought, Immanuel Kant (Ponting 1991). The basis of Kantian philosophy was to treat nature as non-divine, giving humanity a spiritual freedom and domination over it.

The application of Cartesian philosophy to practice implies that nature lacks an intrinsic value, possessing only an instrumental one, thus logically humans have no obligation of duty towards how nature is used. A dramatic example of this philosophy applied to the context of tourism is General Franco's policy for tourism development in Spain in 1959. Entitled the *Plan Nacional de Estabilization*, it possessed an inherent creed of 'crecimiento al cualquier precio' or 'growth at any price'. Similarly, the development of Cancun in Mexico from a village housing 12 Maya families in the 1970s to a resort receiving over 2.5 million visitors per annum, resulting in the destruction of rainforests and mangroves, the filling of lagoons and the levelling of sand dunes (Lynas 2003), is representative of the application of this ethic in tourism.

Whilst Descartes and Kant can be seen to have been influential to attitudes towards nature, not all philosophers subscribed to such a dualistic narrative. Expressing a view of a spiritual connectedness between humans and other species, Benedict Spinoza believed in 'animism' or 'organicism', that a single and continuous force permeated all beings and things, a manifestation of God-created substance. Likewise, Henry Moore, an important philosopher of the seventeenth century, believed that the spirit of God or *anima mundi* was present in every part of nature (Holden 2008).

In more contemporary times, although concerns for the effects of industrial development were expressed by the Romantic movement in nineteenth century Europe, and the effects of capitalism by Henry Thoreau and John Muir during the same century in America, it was Aldo Leopold who overtly expressed the human–nature relationship as an ethical issue in his seminal *A Sand County Almanac* (1949). In his call for a 'land ethic', he observes: 'In short, a land ethic changes the role of *Homo sapiens* from conqueror of the land-community to plain member and citizen of it. It implies respect for his fellow-members, and also respect for the community as such' (1949: 219). Leopold was suspicious of recreation and tourism, viewing it as having a negative effect on wildlife, whilst he perceived the travel trade as encouraging access to nature in bulk and reducing opportunities for solitude.

Concerns over the natural environment and the moral standing of nature have led to a large array of views on environmental ethics that cannot be categorized easily (Booth 1998). Based upon the 'rights' of nature to an existence, Simmons (1993) deduced two broad meanings of the application of ethics to the environment. These are an ethic for the 'use of the environment', which may be summarized as adopting an anthropocentric viewpoint of how we decide to use the resources of the earth, and an ethic 'of the environment', in which all non-human beings are given the same moral standing as that of the human species.

The first of these interpretations is probably the one most people, including decision-makers, would relate to and base their judgements upon. For example, the emphasis of sustainable development and sustainable tourism is very much upon environmental conservation for the well-being of future generations of humans (Cater 2006, Saarinen 2006), rather than concerns of the intrinsic right of nature to an existence. The implication of Simmons' (1993) second position is that if other sentient and non-sentient beings have the same moral standing as humans, they should also have equal rights to those of the human world. Applied to law, this would mean, for example, that a hotel owner could be sued on behalf of the species occupying a coral reef if sewage emissions from the hotel were harming it. If the coral reef species won, the hotel would have to find another means of sewage disposal, as the rights to existence would be paramount.

There is evidence that an ethical viewpoint is influencing the action of tourism stakeholders on the basis that some existing patterns of behaviour can harm the environment. Notably, codes of conduct have been developed by various organisations in the private and public sectors to help guide appropriate tourist behaviour (Holden 2005). Garrod and Fennell (2004), for example, comment that there has been the development of a plethora of codes of conduct for whale-watching over recent years. The private sector seems to be placing emphasis upon demonstrating corporate social responsibility and assessing the performance of companies according to the 'triple bottom line' (Elkington 1998), considering their social, environmental and economic impacts.

The extent that an environmental ethic has or has not permeated into the consumer market for tourism is debatable. Given the growth of green and responsible consumerism it would seem logical to expect evidence of this trend within the tourism market. Certainly, the advent of prefixes such as eco-, sustainable and responsible tourism over the last twenty years would seem to be indicative of such a trend. However, the extent of the willingness of consumers to take adaptive measures to mitigate the negative environmental impacts of tourism, for example by paying higher prices to support environmental conservation schemes or flying less, will only become evident with time and the monitoring of environmental trends within the market.

Conclusion

The relationship that exists between tourism and the environment is a complex one to evaluate because of the range of ethics, meanings and spatial dimensions that are inherent to it. The emphasis of much of what has been written about this relationship, both in academia and the media, tends to assume that any damage to nature as a consequence of tourism activity should be regarded as harmful. Conversely, any use of tourism to ensure nature conservation should be regarded as positive and beneficial. Yet, this dichotomy of the interaction of tourism with nature is in many cases a false one, as the balance of this interaction is not a simple case of good or bad. For example, for tourism to act as an agent of conservation and to alleviate poverty in developing countries, usually necessitates the movement of tourists from developed countries by means of air travel, with the consequent contribution of aircraft emissions to the process of global warming. Similarly, the establishment of protected areas for wildlife conservation, which tourists may choose to visit, may entail the forceful separation of indigenous peoples from their land and natural resources essential to their livelihoods, as with the case of the Maasai in Kenya (Holden 2008).

In a global system that encompasses different cultural and ethical perspectives, it is unsurprising that there will be a range of different viewpoints on the benefits or costs of the interaction between tourism and nature. These differences incorporate the construction of the meaning of 'environment', at its most dualistic ranging from an external reality that can be scientifically theorized or measured, to an entity possessing spiritual connections with other sentient and non-sentient beings, connected to our past and future.

Our cultural constructions of nature will subsequently influence the range of values and sense of duty we attach to nature, determining our perceptions of the levels of acceptable change in the environment for the purpose of tourism development. For example, in emphasizing a spiritual connection with the environment, the first concern of the First Haudenosaunee chiefs in North America when making decisions about the use of nature is to respect the needs of the 'seventh generation' (Holden 2008). This refers to the seventh generation of the past, who sacrificed their own needs to ensure that natural resources were available for the present generation, alongside the seventh generation of the future. Consequently, the First Haudenosaunee may hold different views about tourism development than the non-Haudenosaunee person. Similarly, the traditions of Mayan culture believe that every human has an animal counterpart and vice-versa, thus to harm one is to harm the other.

Our views on the meaning and significance of the environment are not likely to remain static, however, changing with time. Our existing preconceptions may be challenged by several factors, including: scientific discovery, such as global warming; philosophical reflection and debate, such as other sentient beings possessing emotions and consequently having 'rights'; or through new spiritual awakenings, such as the re-discovery of pagan belief systems of nature. Such paradigm shifts are likely to result in a re-evaluation of our ethical relationship with nature. At present, there exists no universal agreement about the environmental ethic of tourism's interaction with nature. Holidays, for example, are advertised that encourage tourists to participate in seal-culling, the killing of baby seals by club or bullet, in Newfoundland and Norway (Jowit and Soldal 2004). From an ethical perspective, it may be argued that whilst the individual right of the seals to exist is being violated, the killing of seals is justified in an eco-holistic framework to maintain fish stocks and the viability of the ecosystem. Of course, how the seals are killed, the enjoyment individuals may attain from it and the financial profits that are attributed from it, add additional complications to this argument.

However, the evaluation of the balance of the interaction of tourism with nature, as either positive or negative, will influence tourism policy and the dynamics of the market. Presently, the dominant paradigm of tourism development is sustainability, supported by a conservation ethic, derived from a realisation that our future well-being and survival is dependent upon the surrounding natural environment. This is a significant shift from the instrumental use of nature that characterised tourism development for most of the twentieth century. The next environmental paradigm shift may be centred upon a spiritual connection to our surroundings, leading to recognition of the rights of nature, and having legal and economic implications for tourism development. Subsequently, the relationship of tourism with nature may be viewed as one of transition, reflecting the changing environmental values of society.

References

Adams, D. (2007) Flights reach record levels despite warnings over climate change. *The Guardian*, 9 May, 3.

Becheri, E. (1991) Rimini and Co – the End of a Legend? Dealing with the Algae Effect. *Tourism Management*, 12, 229-235.

Boo, E. (1990) *Ecotourism: The Potentials and Pitfalls* (Vol. 1). World Wide Fund for Nature, Washington.

Booth, D.E. (1998) *The Environmental Consequences of Growth*. Routledge, London.

Briguglio, L. and Briguglio, M. (1996) Sustainable Tourism in the Maltese Isles. In L. Briguglio, R. Butler, D. Harrison and W.L. Filho (eds) *Sustainable Tourism in Islands and Small States: Case Studies*. Pinter, London, pp. 161–179.

Burac, M. (1996) Tourism and Environment in Guadeloupe and Martinique. In L. Briguglio, R. Butler, D. Harrison and W.L. Filho (eds) *Sustainable Tourism in Islands and Small States: Case Studies*. Pinter, London, pp. 63–74.

Cater, E. (2006) Ecotourism as a Western Construct. *Journal of Ecotourism*, 5, 23–39.

Dubois, G. and Ceron, J.P. (2006) Tourism/Leisure Greenhouse Gas Emission forecasts for 2050: Factors for change in France. *Journal of Sustainable Tourism*, 14, 171–191.

Elkington, J. (1998) *Cannibals with Forks: The Triple Bottom Line of 21st Century Business*. New Society Publishers, London.

Font, X., Cochrane, J. and Tapper, R. (2005) *Pay per Nature View: Understanding Tourism Revenues for Effective Management Plans*. World Wide Fund for Nature, Netherlands.

Garman, J. (2006) If I were … Aviation Minister. *The Ecologist*, December, 23–24.

Garrod, B. and Fennell, D.A. (2004) An analysis of whalewatching codes of conduct. *Annals of Tourism Research*, 31, 334–352.

Gössling, S. and Peeters, P. (2007) 'It does not harm the environment!' An analysis of industry discourses on tourism, air travel and the environment. *Journal of Sustainable Tourism*, 15, 402–417.

Goudie, A. and Viles, H. (1997) *The Earth Transformed: An Introduction to Human Impacts on the Environment*. Blackwell, Oxford.

Hardin, G. (1968) Tragedy of the Commons. *Science*, 162, 1234–1238.
Hawkins, J.P. and Roberts, C.M. (1993) Effects of recreational scuba diving on coral reefs: trampling on reef-flat communities. *Journal of Applied Ecology*, 30, 25–30.
Hill, J.L. and Hill, R.A. (2001) Why are tropical rain forests so species rich? Classifying, reviewing and evaluating theories. *Progress in Physical Geography*, 25, 326–354.
Holden, A. (2005) *Tourism Studies and the Social Sciences*. Routledge, London.
Holden, A. (2008) *Environment and Tourism* (2nd edition). Routledge, London.
Holden, A. and Ewen, M. (2004) Issues of natural resources, social well-being and tourism in the Annapurna area of Nepal. *ASEAN Journal on Hospitality and Tourism*, 3, 51–63.
Huntly, P.M., Van Noort, S. and Hamer, M. (2005) Giving increased value to invertebrates through ecotourism. *South African Journal of Wildlife Research*, 35, 53–62.
IATA (2007) *It's growing. But it will still be small in 2050*. Available at: http://www.iata.org/whatwedo/environment/campaign/bonsai [accessed: 11 July 2007].
IPCC (2007) *Climate Change 2007: The Physical Science* (Basic summary for policymakers). International Panel on Climate Change, Geneva.
Jameson, S.C., Ammar, M.S.A., Saadalla, E., Mostafa, H.M. and Riegl, B. (2007) A quantitative ecological assessment of diving sites in the Egyptian Red Sea during a period of severe anchor damage: a baseline for restoration and sustainable tourism management. *Journal of Sustainable Tourism*, 15, 309–323.
Jenner, P. and Smith, C. (1992) *The Tourism Industry and the Environment*. The Economist Intelligence Unit, London.
Jowit, J. and Soldal, H. (2004) It's the new sport for tourists: killing baby seals. *The Observer*, 3 October, 3.
Kirkby, S.J. (1996) Recreation and the quality of Spanish coastal waters. In M. Barke, J. Towner and M.T. Newton (eds) *Tourism in Spain: Critical Issues*. CABI Publishers, Wallingford, pp. 190–211.
Leopold, A. (1949) *A Sand County Almanac*. Oxford University Press, Oxford.
Lynas, M. (2003) Should I stay or should I go? *The Geographical*, December, 100–104.
McCarthy, M. (2007) Green groups dismayed as flights soar to record high. *Observer*, 8 May, 11.
Mieczkowski, Z. (1995) *Environmental Issues of Tourism and Recreation*. University Press of America, Lanham, MD.
Mishan, E.J. (1969) *The Costs of Economic Growth*. Penguin, Harmondsworth.
Ostrom, E., Burger, J., Field, C., Norgaard, R., and D. Policansky (1999) Revisiting the Commons: Local Lessons, Global Challenges. *Science*, 284, 278–282.

Palaganas, V.P. (1991) Anchor damage on the coral reef of Sombrero Island, Batangas, Philippines. *Coastal Zone: Proceedings of the Symposium on Coastal and Ocean Management*, 4, 3318–3329.

Pearce, D. (1993) *Economic Values and the Natural World*. Earthscan Publications, London.

Ponting, C. (1991) *A Green History of the World*. Sinclair-Stevenson, London.

Rolston III, H. (1988) *Environmental Ethics: Duties to and Values in the Natural World*. Temple Press, Philadelphia.

Rouphael, A.B. and Inglis, G.J. (1997) Impacts of recreational SCUBA diving at sites with different reef topographies. *Biological Conservation*, 82, 329–336.

Saarinen, J. (2006) Traditions of Sustainability in Tourism Studies. *Annals of Tourism Research*, 33, 1121–1140.

Shackley, M. (1996) *Wildlife Tourism*. International Thomson Business Press, London.

Shaw, S. (1993) *Transport: Strategy and Policy*. Blackwell, Oxford.

Simmons, I.G. (1993) *Interpreting Nature: Cultural Constructions of the Environment*. Routledge, London.

Simons, P. (1988) Apres ski le deluge. *New Scientist*, 1, 46–49.

Smith, L. (2007) Eco-tourists save forest jewels from bulldozers. *The Times*, 29 October, 29.

Stone, C.D. (1993) *The Gnat Is Older than Man: Global Environment and Human Agenda*. Princeton University Press, Princeton.

World Tourism Organisation (WTO) (2003) *Climate Change and Tourism*. World Tourism Organisation, Madrid.

Chapter 3

Thinking Globally about Ecotourism Impact: The Contribution of Ecological Footprint Analysis

Colin Hunter

Introduction

Using Ecological Footprint analysis of simplified ecotourism scenarios, this chapter addresses the need to look beyond the immediate destination area when considering the environmental impact and sustainability of ecotourism, highlighting international air travel as an area worthy of particular focus. The chapter draws heavily on Hunter and Shaw (2006, 2007), and the reader is referred to these sources for more detailed explanation of Ecological Footprint analysis procedures and the ecotourism scenarios adopted.

Ecotourism is now apparently one of the fastest growing segments of the global tourism industry (Watkin 2003, Nyaupayne and Thapa 2004), and the popularity of ecotourism holidays has fuelled research interest in ecotourism, generating a substantial body of academic and policy literature (Fennell 2002a). Claims that ecotourism is pioneering the sustainable development of the tourism industry (for example de Villiers 2003) have thus come under growing scrutiny. Although environmental conservation is widely accepted as a key goal of ecotourism activity (for example Ross and Wall 1999, Buckley 2003a), some recent critiques of ecotourism (Mowforth and Munt 2002, Simmons and Becken 2004) have built upon earlier concerns about its frequent reliance on long-haul air travel (Wall 1997).

The global climate change potential of aviation is certainly regarded as a major concern. Current and projected growth rates in air transport, and the uniquely large potential impact of greenhouse gases other than carbon dioxide emitted by aircraft engines into the upper atmosphere, recently led Chapman (2007), for example, to conclude that aviation is currently the most unsustainable mode of transport available. Recent work by Peeters et al. (2007) on tourism transport by EU citizens found that tourism provides the main motive for air transport, with long-haul flights causing a disproportionately high impact in terms of volume of greenhouse gas emissions. Indeed, Simmons and Becken (2004) estimated the carbon dioxide emissions of even a relatively short one-way international flight between Australia and New Zealand to be approximately equivalent to the total transport-related carbon dioxide emissions of a hypothetical 20 day self-drive ecotour in New Zealand.

It may be, therefore, that even if there exists a strong environmental conservation ethic locally at many (even most) individual ecotourism destinations, and that this manifests itself in genuinely low impact ecotourism products, the environmental costs of getting to these destinations is such that the ecotourism sector exerts a significant impact on natural resource use and environmental change at the global scale. In order to test this hypothesis, an environmental impact indicator or metric is required that can express different types of impact generated at different stages in the life-cycle of a tourism product (specifically comparing transit-related with destination-related impact) in the same unit. The Ecological Footprint (EF) provides such an indicator.

Ecological Footprint Analysis and Sustainable Tourism

First described in the early to mid 1990s (Rees 1992, Wackernagel and Rees 1996), the EF is a measure of demands upon the biological productivity and assimilative capacity of the biosphere imposed by a given human population over a certain time period (usually a year). The unique attribute of EF analysis is the expression of demand for all impact components as an equivalent land/sea area (in global hectares, or gha), thereby (it is claimed) facilitating comprehension of environmental impact and providing a powerful educational tool (for example Chambers et al. 2000, Wackernagel and Yount 2000). The EF itself does not exist in real space, but rather can be viewed as the aggregation of myriad actual land and sea areas around the world appropriated by a given population relying on the global movement of raw materials and products. Detailed descriptions of the procedures involved in EF analysis are found elsewhere (for example Wackernagel et al. 1999, Chambers et al. 2000). Typically, calculations account for and combine the use of energy (direct and embodied), foodstuffs, raw materials and water, and also capture transport-related impacts, the production of wastes (including carbon dioxide from the burning of fossil fuels), and the loss of productive land associated with buildings, roads and other aspects of the built environment. A much used benchmark for comparison in EF studies is the so-called 'fair earthshare' value; that is the global average area of productive land/sea space available annually on a per capita basis. One recent estimate of this, which excludes land set aside for non-human species, is 1.8 gha per year (World Wildlife Fund et al. 2004). Other cited values are slightly higher (Chambers et al. 2000, 2004), with around 2 gha per year regarded as a reasonable estimate (Venetoulis et al. 2004). It is important to appreciate that most advocates of the EF technique stress that EF calculations provide conservative estimates of global environmental impact (excluding, for example, the effects of toxic substances), and the tradition in EF analysis is to consciously err on the side of caution when making estimates of resource use and waste production for use in calculations (for example Wackernagel and Rees 1996, Nijkamp et al. 2004).

The potential contribution of EF analysis as an indicator of sustainable tourism has been recognized (Hunter 2002). To date, however, tourism application of EF analysis has been very limited and largely focused on mass tourism. Cole and Sinclair (2002) used EF analysis to assess the sustainability implications of the advent of mass tourism to Manali in the Western Himalayas of India in the 1980s. They found that the EF of Manali increased by over 450 per cent between 1971 and 1995, with the tourism centre becoming increasingly dependent on ecosystem services external to the immediate area. This finding occurred despite international air travel by tourists not apparently being included specifically in the calculation. Gössling et al. (2002) estimated a leisure tourism EF for the Seychelles and found the per-tourist EF to be some 1.9 gha per year, with an average holiday in the Seychelles corresponding to between 17 per cent and 37 per cent of the annual EF of a citizen of an industrialized country. Well over 90 per cent of the total EF was found to be due to air travel to and from the destination. Work by the World Wildlife Fund UK (2002) presented EF analyses of two typical, two-week UK package holiday products from London to the popular Mediterranean destinations of Majorca and Cyprus. For Majorca, the total EF per tourist was found to be 0.37 gha, while the corresponding value for Cyprus was 0.93 gha. Accounting for approximately 50 per cent of the total EF in both cases, air travel was by far the largest single component of the holiday EF (albeit a much smaller proportion than that reported by Gössling et al. (2002), given the relatively short flights involved to the Mediterranean area). More recently, Peeters and Schouten (2006) estimated the EF associated with inbound tourism to Amsterdam for 2001. Transport components were found to dominate the overall EF estimate, accounting for some 70 per cent of the total, with air transport responsible for 87 per cent of the transport EF sub-total. The authors concluded that large reductions in the overall EF of inbound tourism to Amsterdam could be achieved by switching marketing effort from long-haul markets to short-haul markets, given the greater potential for modal shift away from air travel associated with the latter.

These examples illustrate the potential benefit of adopting the EF as a key environmental indicator of sustainable tourism: the EF provides a means of identifying and understanding globally-expressed demands on the biosphere brought about by tourism activity. As the studies demonstrate, this attribute of EF analysis may be particularly pertinent in accounting for the impact of tourism-related air travel to and from the destination. However, it would appear that in these studies the calculated tourism EF was interpreted as a wholly additional 'burden' on the global biosphere. In reality, whilst on holiday (and as recognized by the World Wildlife Fund UK in their study), the tourist is clearly not producing at home the footprint that would normally be created over the same period. The key indicator for any tourism product should therefore be the net, rather than the gross, EF generated, not least because the net estimate is more in keeping with the tradition in EF analysis of consciously erring on the side of caution when estimating the magnitude of a particular activity or group (for example Wackernagel and Rees 1996, Monfreda et al. 2004, Nijkamp et al. 2004). The distinction between gross and net EF estimates may appear rather obvious, and in some scenarios the EF that would have been generated at home will be small compared to the gross holiday EF, meaning that the difference between gross and net EF values is insignificant. Yet, this will not always be so. In the case of ecotourism holiday products it may be particularly important to distinguish between net and gross EF estimates in assessing their apparent sustainability.

Ecotourists generally originate from developed countries where the average per capita EF is high, but frequently holiday in less developed countries (Gössling 1999) with considerably lower average per capita EFs (World Wildlife Fund et al. 2004). Theoretically, therefore, the potential exists for some international ecotourism holiday products to reduce, in net terms, the per capita EF of an individual whilst on holiday compared with the normal EF that would have been generated at home. The need to consider the EF of an international ecotourist in the context of the EF generated by her or him at home has been recognized by Fennell (2002b). Of course, the potential for a net EF reduction assumes that the tourist generates an EF whilst on holiday similar to the average per capita EF of the host population. Clearly, ecotourism activities may be relatively luxurious and resource intensive at the destination (Page and Dowling 2002). However, and in contrast to other types of tourism, products designed to be low impact at the destination occur frequently in the ecotourism sector (for example World Tourism Organization 2003). Thus, with ecotourism holidays in particular, there is at least the potential that net EF estimates will be substantially lower than gross EF estimates, as the rate of demand on natural resources at the destination compares favourably with that at home (as indicated by a large difference between home and destination EF values). Potentially, this difference may be large enough to outweigh the EF generated in transit (even by flying) between home and destination, giving an overall saving on natural resource demand associated with the ecotourism holiday.

Methodological Approach

EF analyses using primary data collected on tourist consumption of natural resources represents a very daunting challenge (Hunter 2002), and hundreds of EF analyses may be required in order to establish a range of EFs for ecotourism products. Using conservative and simplified hypothetical ecotourism scenarios, however, it is possible to establish an indicative range of estimates for the potential minimum net EF of ecotourism, thereby allowing tentative conclusions to be drawn about the likely global impact of ecotourism on natural resources. The findings reported below are not based on real ecotourism products. Rather, using conservative assumptions, the aim was to better understand the potential impact of ecotourism in global terms using hypothetical scenarios of relevance to the ecotourism sector. The ecotourism scenarios adopted paired leading source

Table 3.1 Average per capita Ecological Footprints of source and host countries (2001), and international tourist arrivals for host countries

Country	Ecological Footprint (gha)	Tourist Arrivals (2004)
Source countries		
France	5.8	-
Germany	4.8	-
Japan	4.3	-
United Kingdom	5.4	-
United States	9.5	-
Host countries		
Botswana	1.3	1,268,600
Brazil	2.2	3,797,200
Bulgaria	2.7	3,807,000
Costa Rica	2.1	1,237,000
Jordan	1.9	1,796,000
Kenya	0.9	874,800
Mexico	2.5	20,237,400
Mongolia	0.6	198,400
Morocco	0.9	4,929,000
Nepal	1.9	207,600
Peru	0.9	927,400
Phillipines	1.2	1,686,000
Senegal	1.2	457,400
Thailand	1.6	12,432,600

Source: Hunter and Shaw (2006)

countries involved in generating international tourists with geographically widespread host countries, ensuring variety in net EF estimates. Potential net EF estimates were made using readily accessible secondary data sources and an easily adopted calculation procedure as key attributes of a suitable indicator (Chambers et al. 2000, Bell and Morse 2003).

Five source countries were chosen, these being the leading spending nations on international tourism (USA, Germany, UK, Japan and France) according to the World Tourism Organization (2004). For the USA, two departure cities/airports

Table 3.2 One-way flight distances (km) between source and host countries

	France (Paris)	Germany (Frankfurt-am-Main)	Japan (Tokyo)	United Kingdom (London)	United States (New York)	United States (Los Angeles)
Botswana (Gaborone)	9,021	8,987	18,605	9,362	14,896	18,114
Brazil (Manaus)	11,243	11,630	21,256	10,978	5,656	7,632
Bulgaria (Sofia)	1,750	1,369	10,754	2,039	7,573	10,792
Costa Rica (San Jose)	9,171	9,958	13,179	8,906	3,556	4,878
Jordan (Amman)	3,358	3,022	9,116	3,661	9,195	12,414
Kenya (Nairobi)	6,397	6,462	15,981	6,838	12,372	15,590
Mexico (Mexico City)	9,195	9,453	11,245	8,897	3,365	1,349
Mongolia (Ulaanbataar)	9,345	9,793	3,058	9,310	14,844	11,677
Morocco (Marrakech)	2,126	2,574	11,827	2,293	7,827	11,046
Nepal (Kathmandu)	7,373	7,432	6,719	7,541	13,075	16,293
Peru (Lima)	11,600	11,987	15,755	11,335	5,876	6,736
Philippines (Manila)	10,738	10,290	3,050	10,944	14,190	12,801
Senegal (Dakar)	4,219	4,567	13,803	4,564	6,113	10,082
Thailand (Bangkok)	9,412	8,966	4,641	9,543	15,077	13,275

Source: Hunter and Shaw (2006)

Table 3.3 Procedure for calculating potential net Ecological Footprint estimates (gha per tourist per year)

	Transit zone
1.	Determine the total, round trip, flight distance (km).
2.	Obtain energy use per tourist (megaJoules, MJ) by multiplying flight distance by an energy intensity conversion factor of 2.0 MJ/pkm.
3.	Obtain the equivalent land area (ha of forest) per tourist (per year) required to sequester carbon dioxide production by dividing energy use per tourist by 73 GJ/ha (73,000 MJ/ha) (the number of gigaJoules that 1 ha of forest land will sequester, in carbon dioxide equivalent, per year when liquid fossil fuel is combusted) (World Wildlife Fund et al. 2000).
4.	Allow for the additional radiative forcing of aircraft emissions other than carbon dioxide emitted at altitude (Schumann 1994) by multiplying by a factor of 2.7 (International Panel on Climate Change 1999), giving a new estimate of required forest land (ha).
5.	Multiply by the appropriate 'equivalence factor' (in 2001 this was 1.38) to correct for forest land being more productive than average world space (Chambers et al. 2004), giving a final estimate of the transit zone per tourist footprint in gha per year.
	Destination area
6.	Use the host country average per capita EF as a proxy for the destination area EF of the tourist, reduced pro rata from an annualized value according to the length of stay.
	Net EF
7.	Use the average per capita EF of the source country and the length of stay away from home to calculate the per-tourist EF that would have been generated at home for the period away (again reduced pro rata from an annualized value), and subtract this from the gross per-tourist EF (the sum of steps 1 to 6).

were chosen (New York and Los Angeles) to reflect the size of the country and provide additional geographical diversity to EF estimates. Fourteen host countries were chosen: Botswana, Brazil, Bulgaria, Costa Rica, Jordan, Kenya, Mexico, Mongolia, Morocco, Nepal, Peru, Philippines, Senegal and Thailand. Unfortunately, no global initiative exists for the gathering of ecotourism data (The International Ecotourism Society 2000). Given the consequent lack of information on international ecotourist arrivals around the world, host countries were chosen on the following basis: they represent a geographically diverse group of destinations (at least one from each of the World Tourism Organization's world regions); and, all either provided the World Tourism Organization with information on ecotourism activities during the UN International Year of Ecotourism in 2002, are known to have established authorities to oversee ecotourism activities

(United Nations General Assembly 2003), or are well known as ecotourism destinations through case study material (for example Buckley 2003b). Selected host countries also offered some variety in terms of per capita EF characteristics. Per capita EF estimates for the source and host nations chosen are given in Table 3.1, along with international tourist arrivals for 2004, while flight distances are provided in Table 3.2. Distances were for direct flights where these were available, or assumed reasonably direct connections.

The gross tourism EF can be seen as having two broad components: that generated in the transit zone and that at the destination area. The net tourism EF is simply the sum of the transit and destination area components, less the source country EF for the period away from home. Potentially, there is the complication that a very small footprint will actually be generated by a tourist in the source country even when the tourist is away from home; for example, by leaving on some home heating or security lighting. The normal, major footprint components associated with energy use, transportation, food consumption and the consumption of other raw materials will be absent, however, so it would appear reasonable to assume the home-based footprint to be negligible during the international tourist trip.

The overall procedure for calculating a first order estimate of the potential annual equivalent net per capita (tourist) EF for international ecotourism involving air travel is summarized in Table 3.3. More detailed explanation of aspects of the procedure and the assumptions involved can be found in Hunter and Shaw (2007). The conservative nature of assumptions made implies that estimates should be regarded as potential minima.

Results

Using the above procedure it is possible to estimate the EF of hypothetical ecotourism scenarios using many hundreds of combinations of source and host countries. By way of illustration, the combination of the six source cities (in five countries) and fourteen destination countries chosen (Table 3.1), resulted in 84 net EF estimates being made for a particular length of stay period. Three lengths of stay were used (7 days, 14 days and 21 days) to provide added variety to the chosen source/host country combinations. The 21 day scenario may be particularly long for most ecotourism holidays (for example Page and Dowling 2002), but a long stay increases the likelihood of obtaining small potential net EF values, and it is important to allow for this theoretical possibility. (The reader is reminded that decreasing net EF estimates with increasing length of stay rests on the assumption that the EF generated by the tourist at the destination area is lower than that normally generated by her or him at home).

The US (Los Angeles/Mexico) scenario involved the shortest flight (some 1,349 km, one-way), with the longest one-way flight (at over 21,000 km) between Japan and Brazil (Table 3.2). Transit zone EF estimates (Table 3.4) ranged between

Table 3.4 Transit zone Ecological Footprint estimates (gha per tourist per year) between source and host countries

	France	Germany	Japan	United Kingdom	United States (NY)	United States (LA)
Botswana	1.84	1.83	3.80	1.91	3.04	3.70
Brazil	2.30	2.37	4.34	2.24	1.15	1.56
Bulgaria	0.36	0.29	2.20	0.42	1.55	2.20
Costa Rica	1.87	2.03	2.69	1.82	0.73	1.00
Jordan	0.69	0.62	1.86	0.75	1.88	2.53
Kenya	1.31	1.32	3.26	1.40	2.53	3.18
Mexico	1.88	1.93	2.30	1.82	0.69	0.28
Mongolia	1.91	2.00	0.62	1.90	3.03	2.38
Morocco	0.43	0.53	2.41	0.47	1.60	2.26
Nepal	1.51	1.52	1.37	1.54	2.67	3.33
Peru	2.37	2.45	3.22	2.31	1.20	1.38
Philippines	2.19	2.10	0.62	2.23	2.90	2.61
Senegal	0.86	0.93	2.82	0.93	1.25	2.06
Thailand	1.92	1.83	0.95	1.95	3.03	2.71

Source: Hunter and Shaw (2006)

0.28 gha (Los Angeles/Mexico) and 4.34 gha (Japan/Brazil). Minimum potential net EF estimates were obviously lower (Table 3.5). For a 7 day holiday these ranged between 0.15 gha (Los Angeles/Mexico) and 4.30 gha (Japan/Brazil); the latter value being equivalent to the average annual per capita EF of a Japanese citizen (Table 3.1). For a 14 day holiday, estimates ranged between 0.02 gha (Los Angeles/Mexico) and 4.26 gha (Japan/Brazil). In the former scenario, the holiday potentially accounted for only some 0.21 per cent of the average annual per capita EF of a US citizen. With the longer holiday period of 21 days, one scenario (Los Angeles/Mexico) actually produced a potentially negative net EF finding (-0.13 gha), suggesting the possibility of an overall reduction in demand on global renewable resources in this particular case. This was an isolated result, however, and over one-third (some 36 per cent) of the 14 day holiday scenarios, for example, produced potential net EF values greater than the (annual) 'fair earthshare' value of approximately 2 gha (Table 3.5). In only 20 per cent of 14 day scenarios did the holiday product potentially account for less than the annual average per capita footprint in low-income countries (0.8 gha according to World Wildlife Fund et al. 2004). Overall mean potential net EF estimates for the 7 day, 14 day and 21 day scenarios were 1.76 gha, 1.66 gha and 1.56 gha, respectively, giving a maximum difference of some 11 per cent. The size of the transit component (flight distance) was generally very important in determining the magnitude of potential net EF

Table 3.5 Potential net Ecological Footprint estimates (gha per tourist per year) for source/host country combinations

	France	Germany	Japan	United Kingdom	United States (NY)	United States (LA)
Botswana	1.75	1.76	*3.74*	1.83	*2.88*	*3.54*
	1.67	1.70	*3.69*	1.75	*2.73*	*3.39*
	1.58	1.62	*3.62*	1.67	*2.56*	*3.22*
Brazil	*2.22*	*2.32*	*4.30*	*2.18*	1.01	1.42
	2.16	*2.27*	*4.26*	*2.11*	0.87	1.28
	2.10	*2.22*	*4.22*	*2.06*	0.73	1.14
Bulgaria	0.30	0.24	*2.16*	0.37	1.41	*2.07*
	0.24	0.21	*2.14*	0.31	1.29	1.94
	0.19	0.17	*2.11*	0.27	1.16	1.81
Costa Rica	1.80	1.98	*2.65*	1.76	0.59	0.86
	1.73	1.93	*2.61*	1.69	0.45	0.72
	1.66	1.87	*2.56*	1.63	0.30	0.57
Jordan	0.62	0.57	1.82	0.69	1.74	*2.39*
	0.54	0.51	1.77	0.61	1.59	*2.24*
	0.47	0.45	1.72	0.55	1.44	*2.09*
Kenya	1.21	1.25	*3.20*	1.31	*2.36*	*3.02*
	1.12	1.17	*3.13*	1.22	*2.20*	*2.85*
	1.03	1.09	*3.06*	1.14	*2.03*	*2.68*
Mexico	1.82	1.89	*2.26*	1.76	0.56	0.15
	1.76	1.85	*2.24*	1.71	0.43	0.02
	1.69	1.79	*2.19*	1.65	0.28	-0.13
Mongolia	1.81	1.92	0.55	1.81	*2.86*	*2.21*
	1.76	1.89	0.53	1.76	*2.74*	*2.09*
	1.69	1.83	0.48	1.70	*2.59*	1.94
Morocco	0.34	0.46	*2.35*	0.39	1.44	*2.09*
	0.24	0.38	*2.28*	0.29	1.27	1.93
	0.15	0.30	*2.21*	0.21	1.10	1.76
Nepal	1.43	1.47	1.33	1.48	*2.53*	*3.18*
	1.31	1.36	1.23	1.35	*2.34*	*2.99*
	1.21	1.27	1.15	1.26	*2.15*	*2.81*
Peru	*2.28*	*2.38*	*3.15*	*2.23*	1.04	1.21
	2.18	*2.30*	*3.09*	*2.13*	0.87	1.05
	2.09	*2.22*	*3.02*	*2.05*	0.70	0.88
Philippines	*2.10*	*2.03*	0.56	*2.15*	*2.73*	*2.45*
	2.02	1.97	0.51	*2.07*	*2.59*	*2.30*
	1.93	1.89	0.44	1.99	*2.42*	*2.13*
Senegal	0.77	0.86	*2.76*	0.85	1.09	1.90
	0.69	0.80	*2.71*	0.77	0.94	1.75
	0.60	0.72	*2.64*	0.69	0.77	1.58
Thailand	1.84	1.77	0.90	1.88	*2.93*	*2.56*
	1.76	1.71	0.85	1.80	*2.73*	*2.41*
	1.68	1.64	0.79	1.73	*2.57*	*2.25*

Source: Hunter and Shaw (2006)

Notes: upper value = 7 day holiday, middle value = 14 day holiday, lower value = 21 day holiday; values greater than a 'fair earthshare' of 2 gha shown in *italics*.

UWE Library Services
Frenchay Campus
Tel: 0117 32 82092

Borrowed Items 10/05/2010 15:24
XXXX0940

m Title	Due Date
e economics of money, banking, and financial markets	
00462054	10/05/2010
Ecotourism and environmental sustainability : principles	
00487090	07/06/2010

Amount Outstanding : £4.00

ndicates items borrowed today
w.uwe.ac.uk/library

estimates. This is clear by comparing the findings in Tables 3.2 and 3.4 with potential net EF estimates in Table 3.5. For example, and based on the particular scenarios adopted, the overall mean one-way flight distance to host countries was 11,356 km for Japan, 10,906 km for Los Angeles and 9,544 km for New York. These three source locations were associated with the largest number of scenarios where the potential net EF estimate exceeded a 'fair earthshare' value of 2 gha; 27 scenarios for Japan, 22 for Los Angeles and 18 for New York (Table 3.5). For the other source countries involved (France, Germany and the UK), overall mean one-way flight distance to host countries ranged between 7,496 km (France) and 7,606 km (Germany). For these countries, the number of scenarios where the 'fair earthshare' value was exceeded was much smaller; only seven or eight in each case (Table 3.5).

Discussion

Although these scenario-based estimates must be regarded as indicative at this stage, they do support anecdotal concern (for example Wall 1997) about the possible environmental impacts of flying, and flight distance, associated with increasing ecotourism activity. They also support the findings of previous relevant tourism EF applications in this regard, albeit based upon the use of net, rather than gross, EF estimates and using a variety of scenarios. Even for the 21 day scenarios (Table 3.5), the potential effect of differences between source and host country EF characteristics appeared small in reducing the influence of the transit EF component to overall net EF values. By considering net EF values and by using substantial lengths of stay for scenarios deemed to be relatively low impact at the destination, conservative estimates of relevance to the ecotourism sector have been calculated. Other assumptions in the estimation procedure also generally favoured the generation of conservative estimates, such that the values presented (Table 3.5) may be regarded as potential minima. Despite this, the findings suggest that ecotourism holidays involving air travel are likely to produce an absolute demand on global natural renewable resources. The magnitude of this demand may be very substantial: for a 14 day holiday involving return flight distances of 20,000 km or over, the potential net EF approaches, and then may greatly exceed, 2 gha (an approximation of the annual per capita 'fair earthshare' value).

Furthermore, findings assume that resource use by the ecotourist at the destination is relatively conservative, reflecting that of the host country. Yet, ecotourism activities may occur in much more up-market, resource-demanding contexts with tourists living much more luxurious lifestyles than locals (see, for example, Kontogeorgopoulos 2004). It could be argued, therefore, that in many circumstances, particularly involving popular or 'soft' forms of ecotourism (Page and Dowling 2002), it would be more appropriate to adopt the average per capita EF of the source country as a proxy for the EF generated at the destination (Gössling et al. 2002). Thus, the source country per capita EF would be used in both steps

6 and 7 of the calculation (Table 3.3), and the potential net tourism EF becomes the same as the transit zone EF provided by steps 1 through 5, irrespective of the length of stay. Consequently, the net EF can never be negative, and is simply a function of distance travelled by air. Changing the assumption about the nature of resource demand by ecotourists at the destination in this way produces higher potential net EF estimates, although increases are generally small. Comparing the 14 day findings, for example, in Table 3.5 with those in Table 3.4, mean values (1.66 gha and 1.86 gha, respectively) differ by only some 11 per cent (0.2 gha). Again, this reflects the general dominance of the transit component to overall potential net EF values in the scenarios. Thus, a higher EF at the destination area merely reinforces the key conclusion suggested by these findings: that ecotourism experiences involving international air travel will normally exert an absolute (and substantial) net demand on global natural resources.

Great caution is required, however, in drawing definitive conclusions on the sustainability of ecotourism from the estimates in Tables 3.4 and 3.5. The reader is reminded that findings are not based on real ecotourism products or the use of primary data on tourist natural resource demand, and EF analysis itself carries its own set of assumptions and simplifications (Nijkamp et al. 2004), adding to those made in the rudimentary estimation procedure for the net tourism EF given above. Also, our understanding of the radiative forcing effect of aircraft emissions at altitude remains incomplete (Sausen et al. 2005), and it is not clear how any additional radiative forcing effect should be incorporated into EF calculations. Furthermore, given that ecotourism emerged as an alternative to mass tourism, and much of the debate surrounding ecotourism has been driven by comparison with traditional mass tourism (for example Cater and Lowman 1994), one might base a judgement on the sustainability of ecotourism by comparing this sector with mass tourism. The key question now becomes, is ecotourism more or less demanding of natural resources, in EF terms, than mass tourism? At this stage, without detailed EF analysis for different product types, it is very difficult to provide a concrete answer, particularly as interpretations of ecotourism and ecotourism products may still differ widely (for example Simmons and Becken 2004). But because the ecotourism scenarios (supported by evidence from other studies) suggest a general dominance of the flight-related EF component to net EF estimates, it would only be logical to infer ecotourism as having a greater impact than mass tourism if, at a global scale, ecotourism products generate more air passenger kilometres than mass tourism products.

Unfortunately, there is a dearth of basic information on global ecotourism activity. According to the World Tourism Organisation, in the late 1990s all nature-related forms of tourism may have accounted for some 20 per cent of total international travel (The International Ecotourism Society 2000), but the contribution of ecotourism to this is unknown. It does seem clear, though, that long-haul air travel will increase: worldwide, it is predicted to grow faster, at 5.4 per cent per year over the period 1995 to 2020, than intraregional travel, at 3.8 per cent per year. Consequently, the ratio between intraregional and long-haul air travel may shift from around 82:18 in 1995, to close to 76:24 by 2020 (World Tourism Organization 2004). Should it occur, this increase appears as likely to arise from mass market, package holiday products offered by the large, mainstream operators as from specialized ecotourism products provided by niche operators (for example Meyer 2003). Traditional beach holiday products to long-haul, exotic destinations, for example, are now offered routinely to UK residents (amongst the most frequent long-haul air travellers in the world) by mainstream operators and are increasing in popularity (Meyer 2003). The indirect evidence, therefore, suggests that now and in the near future ecotourism is very unlikely to generate more air passenger kilometres than conventional mass tourism.

Arguably, it is the apparently inexorable growth in international tourism involving air travel that is the fundamental problem, not the growth in one segment of this market. Although the findings presented in this paper suggest that the concerns expressed about ecotourism by some academic researchers may well be warranted on the grounds of transit zone environmental impact, the deeper lesson may be the need for more attention to be paid to transit zone impacts in general, irrespective of the associated product label. It should also be remembered, of course, that genuine ecotourism products carry a commitment at least to the protection of local ecosystems and other desirable actions. As yet, mass tourism products generally do not.

This said, there may be a worrying irony to be found in the assumed relationships between tourism activity, air travel and global climate change that is particularly striking in the case of the ecotourism sector. A focus on the non-consumptive use of local wildlife resources so embedded in principles of ecotourism and wildlife tourism may, even if fully adhered to, miss a larger point and fail to protect local ecosystems from the effects of climate change; effects in part derived from ecotourism-generated air travel itself. By way of illustration, cetacean (whale and dolphin) watching around Scotland's coastal waters has increased in importance in recent years and is ripe for further expansion (MacLellan 1999). Whales and dolphins are Scotland's leading wildlife attraction, providing up to 12 per cent of local income to some remote rural areas (Woods-Ballard et al. 2003). Although tempered with calls for strong local and national protection of cetacean species from the (localized) impacts of tourism activity, a lack of focused marketing effort and low foreign tourist numbers have been seen as important impediments to the more sustainable development of cetacean watching in Scotland. MacLellan (1999), for example, reports work which identified Germany and the USA as particularly noteworthy potential markets, while Woods-Ballard et al. (2003) even suggest manipulating the price structure of tourist trips to incentivize participation by those who travel a long distance. The irony, of course, is that evidence is already emerging of relatively rapid changes to the composition of local cetacean communities, around the north-west of Scotland at least, associated with climate change in the form of ocean warming, such that certain cetaceans may move out of areas of tourist significance, including areas specifically designated for marine conservation (MacLeod et al. 2005).

Conclusions

EF analysis permits a global perspective on tourism-generated natural resource demand using a consistent unit (gha) of evaluation, and it would appear to have much to offer in advancing our understanding of tourism environmental impact. Based on the simplified EF estimates presented above, it can be concluded that any tourism product or activity that relies on air travel, and particularly long-haul air travel, is likely to exert a substantial net demand on natural resource use and contribute to global climate change. This is likely to be the case even where, as with some ecotourism products, the rate of natural resource consumption by the tourist at the destination is markedly less than would be generated by her or him at home. Clearly, however, much remains to be accomplished in the application of EF analysis to tourism, and there is considerable scope for debate over the scenario-based assumptions and findings reported above in the context of ecotourism.

The use of proxy measures, and associated assumptions, is clearly an inherent weakness in the procedure reported in this chapter. Although difficult and time consuming, it is only through the collection and analysis of primary data for real ecotourism products that assumptions can be tested. These data need to include: the consumption of energy (including travel mode and distances), food/beverages, and water; other purchases (for example clothing, gifts); waste products of various kinds; and tourism-related buildings and other forms of infrastructure (Hunter 2002). Approaches to data collection might involve the keeping of diaries by tourists, questionnaire surveys of tourists, and information gathered from hotels, restaurants, bars, and the providers of excursions (Hunter 2002).

Of course, in order to accurately estimate the net EF of any given tourist/holiday, it would also be desirable to gather primary data to determine the normal, household EF of the tourist whilst at home, rather than relying on national average per capita data as a proxy. Household-level EF analysis is, however, in its infancy, although detailed household EF calculators do exist (Hunter et al. 2006), and these might also be adapted to transform resource consumption data from hotels and other tourist businesses into actual per-tourist EF estimates for the destination. As a partial alternative to intensive primary data gathering, one or both of the source and destination country per-tourist EF values could be estimated using more refined secondary data sources. A promising avenue for future research may be to utilize an input–output macroeconomic framework to redistribute and disaggregate national EF information down to the household and tourist economy levels using expenditure data (Wiedmann et al. 2005). Incorporating sensitivity analyses to this type of approach would allow a firmer basis for an understanding of (eco)tourism's natural resource demands, whilst avoiding the difficulties of gathering and utilizing primary data and selecting representative ecotourism products for analysis. Rather than relying wholly on primary data approaches, these might then be more focused on checking and/or calibrating estimates made using secondary data sources. Either way, however, a more concrete and detailed understanding of the EF of (eco)tourism products does require the development of primary data approaches.

Although the ecotourism EF scenarios reported in this chapter suggest that most ecotourism products will exert a net demand on natural resource use and contribute to climate change globally, the ecotourism sector, given its current size, does not in all probability generate a total EF or demand on global natural resources as substantial as traditional forms of mass tourism. However, if ecotourism really is to act as a positive exemplar for the rest of the tourism industry, then there is a particular onus on those engaged in planning, delivering and promoting ecotourism products to consider the impact of these in more global terms and account for air travel natural resource demands in assessing their sustainability. More prosaically, natural ecosystems cannot be immune from the pressures exerted locally by global climate change, and therefore ecotourism products around the world must be regarded as at risk from climate change; change brought about in part by travel, particularly air travel, to the ecotourism destination.

References

Bell, S. and Morse, S. (2003) *Measuring Sustainability: Learning from Doing*. Earthscan, London.

Buckley, R. (2003a) Environmental inputs and outputs in ecotourism: geotourism with a triple bottom line? *Journal of Ecotourism*, 2, 76–82.

Buckley, R. (2003b) *Case Studies in Ecotourism*. CABI Publishing, Wallingford.

Cater, E. and Lowman, G. (1994) *Ecotourism: A Sustainable Option?* John Wiley & Sons, Chichester.

Chambers, N., Simmons, C. and Wackernagel, M. (2000) *Sharing Nature's Interest: Ecological Footprints as an Indicator of Sustainability*. Earthscan, London.

Chambers, N., Griffiths, P., Lewis, K. and Jenkin, N. (2004) *Scotland's Footprint: A Resource Flow and Ecological Footprint Analysis of Scotland*. Best Foot Forward, Oxford.

Chapman, L. (2007) Transport and climate change: a review. *Journal of Transport Geography*, 15, 354–367.

Cole, V. and Sinclair, J. (2002) Measuring the ecological footprint of a Himalayan Tourist Centre. *Mountain Research and Development*, 22, 132–141.

de Villiers, D. (2003) Introduction. In World Tourism Organization (ed.) *Sustainable Development of Ecotourism: A Compilation of Good Practices in SMEs*. World Tourism Organization, Madrid, pp. 9–11.

Fennell, D. (2002a) Ecotourism: where we've been; where we're going. *Journal of Ecotourism*, 1, 1–6.

Fennell, D. (2002b) *Ecotourism Programme Planning*. CABI Publishing, Wallingford.

Gössling, S. (1999) Ecotourism: A means to safeguard biodiversity and ecosystem functions? *Ecological Economics*, 29, 303–320.

Gössling, S., Borgstrom Hansson, C., Horstmeier, O. and Saggel, S. (2002) Ecological footprint analysis as a tool to assess tourism sustainability. *Ecological Economics*, 43, 199–211.

Hunter, C. (2002) Sustainable tourism and the touristic ecological footprint. *Environment, Development and Sustainability*, 4, 7–20.

Hunter, C. and Shaw, J. (2006) Applying the ecological footprint to ecotourism scenarios. *Environmental Conservation*, 32, 294–304.

Hunter, C. and Shaw, J. (2007) The ecological footprint as a key indicator of sustainable tourism. *Tourism Management*, 28, 46–57.

Hunter, C., Carmichael, K. and Pangbourne, K. (2006) Household ecological footprinting using a new diary-based data gathering approach. *Local Environment*, 11, 307–327.

International Panel on Climate Change (1999) *Aviation and the Global Atmosphere. A special report of IPCC working groups I and III.* Cambridge University Press, Cambridge.

Kontogeorgopoulos, N. (2004) Ecotourism and mass tourism in southern Thailand: spatial interdependence, structural connections, and staged authenticity. *GeoJournal*, 61, 1–11.

MacLellan, R. (1999) An examination of wildlife tourism as a sustainable form of tourism development in north west Scotland. *International Journal of Tourism Research*, 5, 375–387.

MacLeod, C.D., Bannon, S., Pierce, G., Schweder, C., Learmonth, J., Herman, J. and Reid, R. (2005) Climate change and the cetacean community of north-west Scotland. *Biological Conservation*, 124, 477–483.

Meyer, D. (2003) The UK outbound tour operating industry and implications for pro-poor tourism. *Pro-Poor Tourism Working Paper No. 17*. Overseas Development Institute, London.

Monfreda, C., Wackernagel, M. and Deumling, D. (2004) Establishing national natural capital accounts based on detailed ecological and biological capacity accounts. *Land Use Policy*, 21, 231–246.

Mowforth M. and Munt, M. (2003) *Tourism and Sustainability: Development and New Tourism in the Third World* (2nd edition). Routledge, London.

Nijkamp, P., Rossi, E. and Vindigni, G. (2004) Ecological footprints in plural: a meta-analytic comparison of empirical results. *Regional Studies*, 38, 747–765.

Nyaupane, G.P. and Thapa, B. (2004) Evaluation of ecotourism: a comparative assessment in the Annapurna conservation area project, Nepal. *Journal of Ecotourism*, 3, 20–45.

Page, S. and Dowling, R. (2002) *Ecotourism*. Prentice Hall, Harlow.

Peeters, P. and Schouten, F. (2006) Reducing the ecological footprint of inbound tourism and transport to Amsterdam. *Journal of Sustainable Tourism*, 14, 157–171.

Peeters, P., Szimba, E. and Duijnisveld, M. (2007) Major environmental impacts of European tourist transport. *Journal of Transport Geography*, 15, 83–93.

Rees, W.E. (1992) Ecological footprints and appropriated carrying capacity: what urban economics leaves out. *Environment and Urbanization*, 4, 121–130.

Ross, S. and Wall, G. (1999) Ecotourism: towards congruence between theory and practice. *Tourism Management*, 20, 123–132.

Sausen, R., Isaksen, I., Grewe, V., Hauglustaine D., Lee, D.S., Myhre, G., Köhler, M.O., Pitari, G., Schumann, U., Stordal, F. and Zerefos, C. (2005) Aviation radiative forcing in 2000: an update on IPCC (1999). *Meteorologische Zeitschrift*, 14, 555–561.

Schumann, U. (1994) On the effect of emissions from aircraft engines on the state of the atmosphere. *Annales Geophysicae*, 12, 365–384.

Simmons, D.G. and Becken, S. (2004) The cost of getting there: impacts of travel to ecotourism destinations. In R. Buckley (ed.) *Environmental Impacts of Ecotourism*. CABI Publishing, Wallingford, pp. 15–23.

The International Ecotourism Society (TIES) (2000) *Ecotourism Statistical Fact Sheet 2000*. Available at: http://www.ecotourism.org/research/stats/files/stats.pdf.

United Nations General Assembly (2003) *Assessment of the results achieved in realizing aims and objectives of the International Year of Ecotourism (2003)*. Available at: http://www.world–tourism.org/sustainable/IYE/IYE–Rep–UN–GA–2003.pdf.

Venetoulis, J., Chazan, D. and Gaudet, C. (2004) *Ecological Footprint of Nations*. Redefining Progress, Oakland, CA.

Wackernagel, M., Onisto, L., Bello, P., Linareas, A.C., Falfan, I., Garcia, J., Guerrero, A. and Guerrero, C. (1999) National natural capital accounting with the ecological footprint concept. *Ecological Economics*, 29, 375–390.

Wackernagel, M. and Rees, W.E. (1996) *Our Ecological Footprints: Reducing Human Impact on the Earth*. New Society Publishers, Gabriola Island, BC.

Wackernagel, M. and Yount, J.D. (2000) Footprints for sustainability: the next steps. *Environment, Development and Sustainability*, 2, 21–42.

Wall, G. (1997) Is ecotourism sustainable? *Environmental Management*, 21, 483–491.

Watkin, J. (2003) The Evolution of Ecotourism in East Africa: From an Idea to an Industry. *IIED Wildlife and Development Series No. 15*. International Institute for Environment and Development, London.

Wiedmann, T., Minx, J., Barrett, J. and Wackernagel, M. (2005) Allocating ecological footprints to final consumption categories with input–output analysis. *Ecological Economics*, 56, 28–48.

Woods-Ballard, A.J., Parsons, E.C.M., Hughes, A., Velander, K., Ladle, R. and Warburton, C.A. (2003) The sustainability of whale-watching in Scotland. *Journal of Sustainable Tourism*, 11, 40–55.

World Tourism Organization (2003) *Sustainable Development of Ecotourism: A Compilation of Good Practice in SMEs*. World Tourism Organization, Madrid.

World Tourism Organization (2004) *Tourism Highlights: 2004*. World Tourism Organization, Madrid.

World Wildlife Fund, United Nations Environment Programme, World Conservation Monitoring Centre, Redefining Progress and The Centre for Sustainability Studies (2000) *Living Planet Report 2000*. World Wildlife Fund, Gland, Switzerland.

World Wildlife Fund, United Nations Environment Programme, World Conservation Monitoring Centre and Global Footprint Network (2004) *Living Planet Report 2004*. World Wildlife Fund, Gland, Switzerland.

World Wildlife Fund-UK (2002) *Holiday Footprinting: A Practical Tool for Responsible Tourism*. World Wildlife Fund–UK, Godalming.

PART 2
Thematic Case Studies

Chapter 4

A Comparative Study of Tourism Impacts on Alpine Ecosystems in the Sagarmatha (Mt. Everest) National Park, Nepal and the Huascarán National Park, Peru

Alton Byers

Introduction

This chapter presents two case studies of contemporary landscape change in the alpine zones of Sagarmatha (Mt. Everest) National Park, Khumbu, Nepal and the Huascarán National Park, Cordillera Blanca, Peru. Research methods employed here include systematic belted line transects, random plot sampling, repeat photography and oral testimony of local residents. Results suggest that the comparatively neglected alpine zones within both parks have been significantly degraded during the past 25 to 40 years as a result of poorly controlled adventure tourism. Impacts within the alpine zone include the over-harvesting of fragile alpine shrubs and plants to provide fuel for expeditions and tourist lodges, overgrazing, accelerated hillslope erosion, uncontrolled lodge building, health hazards related to poor sanitary practices, and improper garbage disposal. The chapter stresses the importance of integrated, applied research to the clarification of problems, remedial project design and monitoring of project impacts; it also provides a short account of how information generated by the research led to the ongoing implementation of an international programme, designed to conserve, protect and restore fragile alpine environments throughout the mountain world.

Case Study 1: Sagarmatha (Mt. Everest) National Park, Nepal

Physical Context

The Sagarmatha National Park, officially created in 1976, is located in northeastern Nepal in the Solu Khumbu District of the Sagarmatha Zone (Figure 4.1). This 1,113 km^2 park is roughly triangular in shape and virtually enclosed by mountain peaks in excess of 6,000 m. The park's northern boundary forms the international frontier with the Tibet Autonomous Region of China over a distance of 40 km.

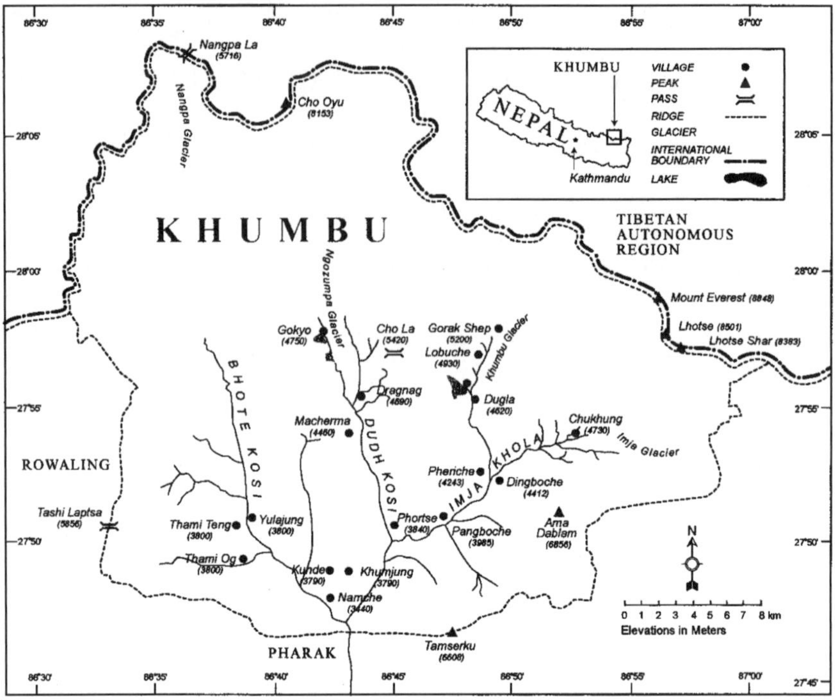

Figure 4.1 **The Sagarmatha National Park, Khumbu, Nepal, highlighting key villages and peaks mentioned in the text**

It includes the world's highest mountain, Mt. Everest (8,848 m), also known to Tibetan speakers by the centuries-old name of Chomolangma and to Nepali speakers as Sagarmatha.

Much of the regional physiography is composed of south- and north-facing slopes. Southern aspects at the lower elevations (3,600 m to 4,030 m) are generally characterized by dry, terracetted shrub/grasslands of *Rhododendron lepidotum*, *Cotoneaster microphyllus*, *Juniperus recurva/indica* and *Berberis sp.* shrubs, with a diverse variety of herbs, forbs and grasses. The more moist and less accessible northern aspects contain large areas of mixed fir (*Abies spectabilis*), silver birch (*Betula utilis*) and rhododendron (*R. campanulatum*, *R. arboreum*, *R. campylocarpum*, *R. hodgsonii*), with locally important associations of maple (*Acer caudatum*), mountain ash (*Sorbus microphyllus*), willow (*Salix eriostachya*, *Salix daltoniana*), bamboo (*Arundinara sp.*) and juniper (*Juniperus recurva*). The alpine zone (4,100m to 5,500 m) is divided into a lower belt dominated by moist alpine scrub of dwarf rhododendrons (*Rhododendron setosum*, *nivale*, *lepidotum* and *anthropogon*) and prostrate junipers (*Juniperus recurva or indica*), and an upper belt dominated by *Kobresia pygmaea* mats and cushion plants such as

Arenaria byophylla, *Anaphalis cavei* and *Leontopodium monocephalum* (Meihe 1987, 1989).

Despite the fact that alpine ecosystems are one of the most biologically diverse habitats in the world per unit area (Körner 1999), plant growth cycles are slow and even modest disturbance can impact upon ecosystem structure and composition for decades after their occurrence.

Geographically, the Khumbu region lies within the subtropical Asian monsoon zone, which is characterized by pronounced summer rainfall maxima, with more than 80 per cent of the annual precipitation falling during a period of approximately four months between June and September (Mani 1981, Barry and Chorley 2003). Topographic barriers to the full force of the monsoon are imposed by a series of mountain ranges along the park's southern borders that include the Numbur, Konge Ri, Tramserku and Kang Taiga Himal. The average annual precipitation decreases with elevational gain; for example, it is 1,148 mm/yr in Namche Bazaar (3,440 m) and 518 mm/yr in Dingboche (4,412 m) (Byers 1987a). Winters are normally dry, although occasional mid-latitude cyclones, driven by the subtropical jet stream that takes its winter position just south of the Himalayan ridge, can cause heavy snowfall events (Zimmermann et al. 1986). Permanent snowline is situated at approximately 6,000 m on south-facing slopes and 5,700 m on north-facing slopes (Haffner 1972).

Tourism Context

Tourism to the region, largely for adventure trekking and mountaineering, has grown significantly since the early 1960s, from 20 visitors in 1964 (Naylor 1970) to 18,200 during the 1997 to 1998 season (Nepal 2000, Nepal et al. 2002). Reportedly, more than 27,000 visitors arrived in 2001 (Sherpa, N.W. 2001, pers. comm.). Most trek to the Tengboche monastery (3,857 m) or Everest base camp region (5,364 m), staying in lodges, teahouses or tented camps of organized treks. Unprecedented international attention was focused on the Khumbu in May of 1996, when a freak snowstorm killed 12 climbers on Everest (Boukreev and DeWalt 1997, Krakauer 1997). Surprisingly, requests for Everest climbing permits increased in the year following the tragedy, as did the number of trekkers who visited the base camp region. The popularity of films such as the 1996 large format IMAX movie *Everest*, and books (for example Coburn 1997), attests to the enduring fascination that much of the world holds for the Khumbu, Everest and its Sherpa inhabitants. The discovery of George Mallory's body on the Tibetan side of Everest in 1999 (Hemmleb et al. 1999) resulted in additional global interest in the region.

Following more than 25 years of dramatic tourism and infrastructure growth (see Nepal 2001 and Nepal et al. 2002 for case studies of structural change within Namche Bazar between 1955 and 1997), new lodges continued to be constructed in nearly every village and seasonal settlement between Lukla and the Everest basecamp throughout the monsoon season of 2001, presumably in anticipation of even greater tourist numbers during the 2001/02 season. Unfortunately, escalating civil unrest within Nepal, combined with the terrorist attacks of 11 September 2001 in the United States, resulted in the cancellation of all climbing and approximately half of the trekking activity in the Khumbu during the 2001 season. Prior to that time, tourists had outnumbered the local Sherpa populations as early as the late 1970s, bringing with them a variety of positive influences (additional income, international exposure and cross-cultural exchange) and negative influences (inflation, the modification of traditional values, changes in the local workforce and loss of life among high altitude Sherpa climbers) (see Coppock 1978, Fürer-Haimendorf 1984, Fisher 1990, Stevens 1993, Lachapelle 1998, Ortner 1999, Nepal 2000, 2001). Because the Khumbu region was relatively unaffected by the Maoist insurgency and its impacts on tourism in Nepal, these numbers began to climb back to their previous levels by 2003 (Sagarmatha National Park 2008).

Contemporary Landscape Change

Five separate research expeditions to the Sagarmatha National Park between 1984 and 2001 provided the opportunity to examine contemporary landscape change within the region. The collective results of these expeditions suggest that most subalpine shrub/grassland and forest landscapes below 4,000 m are relatively stable and that the extent of subalpine forest remains essentially unchanged from the 1950s, in spite of a large body of literature stating otherwise (see Byers 2005: 113–117). In contrast, much of the upper Imja Khola alpine zone above 4,000 m is highly disturbed, particularly those regions that have experienced heavy tourist traffic during the past 25 years.

In the Khumbu, I have suggested that the bulk of contemporary alpine degradation has been directly linked to the recent and significant growth of unregulated adventure tourism, where the alpine zone is either a destination in itself (for trekking groups) or passed through en-route to the higher base camps (by climbing expeditions) (see Stevens 2003, Byers 2005). This position challenges popular, conventional and sometimes donor-driven hypotheses stating that '[increased] income generation from small-scale commercial resource-based enterprises will lead to conservation success' (Margoluis and Salafsky 1998: 107). Rather, the evidence suggests that the significant economic benefits enjoyed in recent years by many Sherpa people, lodge owners and trekking/mountaineering companies utilizing the alpine zone have not been matched by concurrent levels of high altitude land stewardship (Byers 2002, 2005). The relative contributions of disaggregated 'impact groups' – lodges, porters, yak herders, packstock (yak/cattle crossbreed) and expeditions – have yet to be determined. Lodge owners and mountaineering expeditions, however, have most likely been responsible for the bulk of juniper shrub and *Arenaria* harvesting during the past 20 years. Using repeat landscape photographs and systematic belted groundcover transects, I was able to determine that between 40 per cent and 50 per cent of shrub juniper ground cover was lost on Dingboche hillslopes between 1961 and 1995 (Byers 1996,

1997, 2005). Further, alpine soil study plots installed near Dingboche (4,412 m) in 1984 returned high seasonal soil losses of between 20 and 40 tonnes per hectare over the March to November season (Byers 1987a, b). This was linked partly to the harvesting of shrub juniper for fuelwood (servicing high altitude tourist lodges), resultant disruption of the 'geomorphic glue' supplied by the shrubs to the hillslope soil, loss of protective groundcover, and alpine turf removal for lodge and wall construction and grazing. Additionally, key informant interviews with lodge owners, yak herders and park personnel in 2001 (Byers 2005) linked shrub juniper harvesting and hillslope degradation with tourism lodges and traffic throughout the Dingboche Valley over the last three decades. Stacks of cut juniper were a common sight observed near tourist lodges and yak herder camps in 2001 (see also Stevens 1993, 2003, Byers 2005). Porters, left to fend for themselves at the end of a day's trek, were found to be responsible for a large proportion of juniper denudation in order to stay warm at night in whatever shelters they could find, utilizing cut juniper for heating and cooking purposes. In spite of a sound knowledge of grazing systems (March 1977, Alirol 1979, Bjönness 1980, Bishop 1989; Brower 1991, Stevens 1993), it was found that the impacts of traditional yak herders on the Khumbu alpine ecosystem, specifically in terms of fuelwood requirements and other impacts, were not well documented and in need of further study. Likewise, tourist packstock numbers continue to increase with each year and are likely to be responsible for a corridor of trail,[1] hillslope, infrastructure and ecosystem impacts that some consider to be far greater than those imposed by porters (for example, grass to feed the packstock must be imported from the lower altitudes during the tourist season, instead of being purchased locally as in previous decades; and profits from the sale of grass now exceed those of selling potatoes (Sherpa, A.R. 2004, pers. comm.)). Notable infrastructure growth has also occurred in the Khumbu during the past 20 years. This is particularly evident within those villages situated along the main tourist/expedition trails and intersections, for example, Namche (see Nepal 2000, Nepal et al. 2002), those located conveniently between daily destinations (such as Tengboche), or those exhibiting economically 'progressive' attitudes toward growth and tourism (Brower 1991).

Similar alpine degradation scenarios have been documented anecdotally elsewhere in the vicinity of the Everest massif, such as the Makalu basecamp to the east (Byers and Banskota 1992, Carpenter 1993, Cox 1999). In the Mera Peak region southeast of the park, a popular trekking and climbing destination since the mid 1990s, Cox (1999) reported that since 1996 'the dwarf juniper [has been] devastated by trekking groups and local lodges. At each lodge huge piles of firewood are burned in the stove, [and] porters can be seen out daily chopping away at the juniper'. In 2007, I was able not only to verify Cox's observation but, through interviews, belted transects, and direct measurement of cut juniper

1 Interestingly, a 2007 reconnaissance of Kilimanjaro showed that the condition of trails along the popular Marangu route to the summit was very good, and most likely related to the fact that only porters, and not pack animals, are permitted to carry loads.

fuelwood, was also able to estimate that more than 406,350 kg of shrub juniper per year is harvested by lodges and herders, and that more that 50 per cent of the juniper groundcover has been removed during the past two decades (Byers 2007a). A similar scenario on the north (Chinese) side of Mt. Everest is reportedly taking place. Although tourists have only visited the region since 1980, concern about drastic change in land cover was expressed as early as 1989 by Qomolangma Nature Preserve officials (Management Bureau 1989, Sherpa, L.N. 2001, pers. comm.). Dramatic increases in mountaineering expeditions, the over-harvesting of shrub juniper for fuelwood, years of garbage accumulations, and other processes similar to those found on the Nepal Sagarmatha National Park side, were reportedly taking place in upper Rongbuk Valley (Taylor-Ide 2002, pers. comm.), prompting Chinese officials to recently consider the possibility of limiting the numbers of climbers to the region (Ang 2008). While the focus of their planned 2009 expedition is again concerned with cleaning garbage accumulated over 50 years, I continue to maintain that garbage is still largely cosmetic, and that 'one of the tougher challenges is preventing people from ripping the local shrub juniper', which can cause irreparable damage to the fragile alpine ecosystems (Byers, quoted in Ang 2008).

Fortunately, by 2003, the problems of the Khumbu alpine zone began to receive increased acknowledgement in the literature and media, partly as a result of the research reported here (see Klatzel and Byers 2001, Nepal et al. 2002, Chepesuik 2003, Deegan 2003, National Geographic Society 2003, Stevens 2003), and partly from subsequent publications and internet links (for example Byers 2005, Byers 2007b). On 28 May 2003, during the celebrations of the fiftieth anniversary of the climbing of Mt. Everest, the American Alpine Club announced its financial support for a pilot scale 'community-based alpine conservation and restoration project' in the Khumbu (UIAA 2003, AAC 2004), which was subsequently matched by grants from the National Geographic Society's Conservation Trust, US Agency for International Development, United Nations Development Programme, and private donations. The project, entitled the 'Khumbu Alpine Conservation and Restoration Project', is based on the recommendations of local people interviewed during the course of the 2001 field work (Byers 2003) and other long-term studies (Brower 1991, Stevens 1993, 2003) that isolated a desire for participatory approaches, partnerships with local NGOs, project management training opportunities and the development of educational materials targeted at lodge owners, trekking agencies, national park visitors and local schools. The Khumbu Alpine Conservation Committee (KACC), the first of its kind in Nepal, was formed by local people in May 2004 (Sherpa, A.R. 2003, 2004) with technical and financial support from The Mountain Institute and the American Alpine Club. It has since banned the use of juniper for fuelwood, as incense for *puja* (worship) ceremonies performed by expeditions in the Everest basecamp region, and as cover for potato pits (saving an estimated 100,000 kg of shrub juniper per year); renovated a lodge in Chukung that provides porters with a warm place to sleep; established a kerosene depot and KACC headquarters in Dingboche; and published visitor information and environmental brochures. Other activities include the continued training and capacity building of local people, the development of alpine-specific educational materials, the construction of a cattle-proof exclosure designed to demonstrate high altitude hillslope restoration, and training in the production of alternative income generating activities, such as the production of juice for sale to trekkers and climbers from local seabuckthorn berries (Sherpa, A.R. 2008). In 2007, the Argosy Foundation and Patagonia Inc. provided generous support toward the establishment of a unique collaboration between The Mountain Institute and the American Alpine Club, dubbed the Alpine Conservation Partnership, that has since replicated the successful models first developed in the Everest region in the Hinku Valley of Makalu-Barun National Park, Nepal; Ishinca Valley of Huascarán National Park, Cordillera Blanca, Peru; Los Glaciares National Park, Patagonia; and conducted a detailed reconnaissance for Kilimanjaro (Byers 2007c). Phase II (2009-2015) could expand the project even further, with prospective sites including the Wind River Range, Wyoming; Aconcagua National Park, Argentina; Kilimanjaro National Park, Tanzania; Mt. Kenya National Park, Kenya; and Karakoram National Park, Pakistan. In 2007 and 2008, the Khumbu project also made notable contributions to our understanding of contemporary climate change impacts on glaciers using repeat photography (Byers 2007d) that was featured in a special photographic exhibit during the fall/winter of 2008 in selected cities of Europe, South Asia, and North America (see: http://www.changing–landscapes.com/).

Case Study 2: Huascarán National Park, Ancash Department, Peru

Physical and Human Context

Set high in the Ancash Department of the Peruvian Andes, the 3,400 km^2 Huascarán National Park (HNP) is one of the world's premier mountain protected areas (Figure 4.2). Designated a national park in 1975, a UNESCO Biosphere Reserve in 1977 and a World Heritage Site in 1985, HNP encompasses nearly all of the Cordillera Blanca, the highest glaciated mountain range within the tropical zone. 41 glacial valleys (*quebradas*) transect the Cordillera Blanca from both the east and west. Between 3,000 m and 4,800 m, the terrain is characterized by high altitude grasslands (*puna*) with the higher, steeper slopes and ravines containing remnant forests of *queñual* (*Polylepis sp.)* that harbour much of the park's diverse flora and fauna (Fjeldså and Kessler 1996, Tohan 2000). 799 plant, 112 bird and 10 mammal species have been reported (Smith 1988, Brako and Zarucchi 1993, Parkswatch 2005). Other landscape and vegetation features include: remnant *Puya raimondi* stands; *bofedales*, or high altitude wetlands; *matorrales* shublands of *Gynoxis* and *Baccharis sp.*; dense, matt-like formations of *Oreoboluys* and *Calamagrostis sp.*; hundreds of high altitude lakes; and a high Andean alpine vegetation zone of

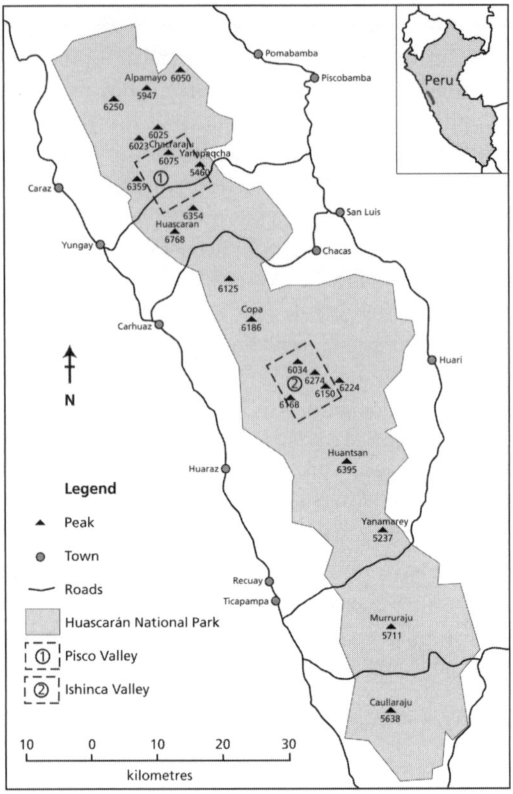

Figure 4.2 The Huascarán National Park, Ancash Department, Peru, highlighting the park boundary and the locations of the Ishinca and Pisco Valleys

low-growing plants adapted to the cold and harsh high altitude climate (Parkswatch 2005).

The park's lower reaches make up much of its 5,710 km^2 buffer zone. This buffer zone is home to an estimated 480,173 residents (INEI 2002), many of whom live in small rural villages under traditional Andean subsistence conditions. Livelihoods in the region have historically revolved around agriculture and livestock, and local residents have long made use of resources within the HNP that include fresh water, pastures for livestock, medicinal plants and fuel wood. Since the formal establishment of the park in 1975, limits have been placed on these resource uses, although communal grazing within park boundaries continues with little or no park input. Each valley entrance is gated and locked, with access controlled carefully by local communities. The obvious presence of cattle and cattle impacts, from the lower sites of interest to the highest basecamps, is of initial surprise to most visitors expecting a more 'pristine' wilderness setting.

During the past decade, HNP has nevertheless attracted rapid growth in both conventional (sightseeing) and adventure (mountaineering, trekking, mountain biking, bungee jumping) tourism. Based on data collected from the Llanganuco and Carpa control posts, close to 100,000 tourists have visited the park each year since the demise of the *Sendero Luminoso* (Shining Path) insurgency in the early 1990s (Byers 2000). In 2004, 76,342 tourists visited the park via these two-entrance stations, 62,725 of whom were Peruvian and 13,617 of whom were foreign (Parkswatch 2005). Because visitors can enter the park through dozens of valleys with no control posts, these totals are most likely underestimates. However, an overall levelling-off of numbers is believed to be occurring (Dourjeanni 2006, pers. comm.).

The communities bordering HNP, especially those on the western side of the range, serve as the primary access points to the national park, and local residents often work seasonally as muleteers, porters, cooks and camp guardians. Providing these auxiliary services to tourists has become the most important source of cash income for many residents, and with this has come increased competition for jobs. In some cases, inter- and intra-community competition and conflict for the benefits from tourism has increased (Recharte, J. 2006, pers. comm.).

The most pressing environmental challenges currently facing HNP, similar to those reported almost a decade ago (see Byers 2000), include the slow but insidious reduction of remaining *queñual* (*Polylepis sp.*) forests; high populations of cattle and overgrazing of alpine and subalpine pastures; garbage and sanitation problems in high altitude basecamp areas; changing and uncertain park policies regarding access to climbing areas; uncertain land titles and park boundaries; conflicts with communities over land ownership and tourism fees; government policies supportive of resource extraction within park boundaries; and subsequent external pressures to develop mining, hydroelectric and new tourism infrastructure (Byers 2000, Bartle 2004).

Recently, the impacts of climate change and global warming within the HNP have received global attention. UNESCO was recently petitioned 'to … list the Huascarán National Park as World Heritage in Danger due to the serious and specific, ascertained and potential dangers from the combined effects of global climate change' (Foro Ecológico del Perú 2006: 2). Other anticipated, although poorly understood, impacts include changes in agricultural patterns (Byers 2007e); the depletion of glacier-fed water supplies for power, agriculture, and drinking water (Bury 2007); increases in conflicts over diminishing water supplies for irrigation (Rhoades et al. 2006); and negative impacts on mountain tourism (see Price and Barry 1997). Despite the accelerated interest in climate change issues, however, little systematic and empirically-grounded work has been done towards better understanding what the actual impacts of climate change are, and will be, on people's lives and livelihoods in the Cordillera Blanca region and elsewhere in the mountain world (see Rhoades et al. 2006).

Concerns regarding contemporary tourism impacts within the region have been focused primarily on litter and the poor sanitary conditions of many base

Table 4.1 Variables measured in field sampling plots in the Ishinca and Pisco Valleys, Huascarán National Park

Altitude	3500–5000	m
Slope	0–100	per cent
Aspect	0–360	degrees
Bare Ground	0–100	per cent
Stipa sp. groundcover	0–100	per cent
Ichu sp. groundcover	0–100	per cent
Stipa/Ichu sp. groundcover	0–100	per cent
Calamagrosis sp. groundcover	0–100	per cent
Moss/Cushion Plant cover	0–100	per cent
Detritus groundcover	0–100	per cent
Herbaceous groundcover	0–100	per cent
Rock groundcover	0–100	per cent
Baccharis/Lupine groundcover	0–100	per cent
Aster sp. groundcover	0–100	per cent
Berberis sp. groundcover	0–100	per cent
Valeria sp. groundcover	0–100	per cent
Tola sp. groundcover	0–100	per cent
Total shrub groundcover	0–100	per cent
Terracettes	0–20	number
Terracette cover	0–3	rank
Cattle impacts	0–3	rank
Turf harvesting	0–3	rank
Burning	0–3	rank
Turf exfoliation	0–3	rank
Dead branches	0–3	number
Mass wasting	0–3	rank
Overland flow	0–3	rank
Cut stems	0–100	number

camp and high camp regions (INRENA 1996, 2003). Less attention has been given to the other potential impacts of tourism on high altitude ecosystems (which have been documented elsewhere; see Byers 2005), including the use of alpine shrubs for fuel in the absence of easily obtainable fuelwood by lodges, porters, trekking and climbing groups; an increase in traffic from pack animals carrying tourist loads, thus increasing grazing and soil compaction problems; trail erosion and 'impact corridors' created by increased numbers of tourists, porters and packstock; removal of protective alpine turf for building materials (such as walls and patio floors of lodges); poaching and overharvesting of medicinal and aromatic plants; and burning of alpine shrubs and hillslopes to expand the grazing area.

The fundamental objectives of the research reported here were:

1. To ascertain the existence and/or respective magnitudes of each of the anticipated impacts on the condition of alpine/high grassland groundcover.
2. To examine the preventative, restorative and management solutions available to help minimize unsustainable land use processes within the HNP's alpine ecosystems.

The research focused, in particular, on the Ishinca and Pisco Valleys within the HNP. The easy access of the Ishinca Valley from Huaraz, its variety of climbing objectives and its tourism support infrastructure make it one of the most popular adventure tourism destinations in the Cordillera Blanca. The valley is located on the western flank of the Cordillera Blanca, 20 km north-east of Huaraz (Figure 4.2). The numbers of climbers visiting the valley each year is unknown, but is probably in the order of several thousand. Likewise, an unknown number of burros, transporting the loads of tourists, are present during the peak trekking and climbing months of June to September. The Pisco Valley is also located on the western flank of the Cordillera Blanca, approximately 75 km north of Huaraz (Figure 4.2). Between 3,000 and 5,000 trekkers and climbers visit the valley each year, along with an estimated several thousand burros left to range freely once the tourist loads are deposited.

The Ishinca and Pisco valleys were chosen as study sites because of their popularity among climbers and trekkers; long histories of grazing activity; ongoing, complementary research being conducted by other researchers (for example Lipton 2002, Tovar and Oscanoa 2002); anecdotal evidence suggesting that cattle and human impacts within the valleys were high (Tohan 2000, French 2003); and potential as future community development/conservation project sites for The Mountain Institute's (TMI) Andean Programmes based in Huaraz.

Field Methods

Field work in the Ishinca Valley was conducted between 26 May and 4 June 2002, and in the Pisco Valley between 4 and 16 June 2002. Each valley was revisited in August 2007 as part of a reconnaissance sponsored by the Alpine Conservation Partnership project, which had been established in the interim (see previous section). Research methods included systematic belted line transects, random plot sampling and oral testimony of local residents. Systematic belted line transects were selected because of their ability to sample evenly across environmental and species gradients. Transects commenced at the valley floor (4,200 m) and ended in the upper alpine zone (4,800 m). Three 25 m^2 sampling quadrats were randomly established at every 50 m altitudinal interval along the ascending transect line (with a total of 207 sampling plots within the two valleys). Data for 40 variables were collected from each 25 m^2 plot, which included cover percentages of bare

Table 4.2 Ranges and mean values of selected variables from the Ishinca and Pisco Valleys, Huascarán National Park

Variable	Unit	Ishinca (n=143)		Pisco (n=63)	
		range	mean	range	mean
Topography					
Altitude	m	4235–4775	4484	4489–4802	4704
Slope	per cent	20–75	57	12–75	51
Aspect	degrees	8–375	207	1–357	92
Groundcover					
Moss	per cent	0–3.3	0	0–12	2
Bare ground	per cent	0–73	24	8–73	41
Grass	per cent	0–80	32	1–80	26
Detritus	per cent	0–5	0	0–20	1
Herbaceous	per cent	2–54	16	0–54	1
Rock	per cent	0–54	26	10–53	25
Shrub	per cent	0–22	3	0–22	4
Disturbance					
Terracettes	number	0–5	2	1–5	3
Terracette cover	rank	0–3	2	1–3	2
Cattle disturbance	rank	0–3	2.2	1–3	2.4
Turf harvesting	rank	0–3	0.1	1–3	0
Cut shrubs	rank	0–3	0	0–1	0
Mass wasting	rank	0–3	1.7	0–3	2.3
Overland flow	rank	0–3	1.6	0–3	2.2

ground, herb, rock, detritus, shrubs, burning, cut shrubs, presence and condition of cattle terracettes and other attributes (Table 4.1). Each stratification also included one 400 m^2 quadrat for detailed plant inventories (with a total of 75 quadrats). Plot positions were recorded with a Garmin Summit GPS. Interviews were conducted with local people, *refugio* staff, *arrieros*, national park officials, trekkers and climbers in an effort to better understand contemporary issues and opportunities from the point of view of the stakeholders in question.

Similar to the Everest region, the Andean alpine has been curiously neglected in the conservation and development literature in spite of its high biological, hydrological, economic and cultural importance (Byers 2005). In the Cordillera Blanca, alpine and subalpine vegetation represent an extremely important source of medicinal plants, fresh water and grazing land. Culturally, these high elevations areas are the gateways to the homes of the apu, or mountain gods. They are also the sites of the many climbing and trekking routes and basecamps that generate substantial revenue, which has become a permanent and vital component of local livelihoods.

Contemporary Landscape Change

Table 4.2 shows averaged values for selected variables from the Ishinca and Pisco Valleys. Of immediate interest are the unusually high percentages of bare ground found in both valleys at a time when a more continuous groundcover would have been expected (at the close of the rainy season). Additionally, the average percentage of bare ground per plot found in the Pisco Valley (41 per cent) is not only extremely high, but is nearly twice that found in the Ishinca Valley (24 per cent). High values for other variables such as rock cover, mass wasting, number of terracettes per plot and overall cattle disturbance suggest that cattle are the primary source of surficial impacts in the high altitude landscape. Collectively, the findings also support the observations of Tohan (2000) and French (2003) regarding the apparently heavier grazing pressures in Pisco. Average values for numbers of terracettes, cattle disturbance, mass wasting and overland flow per plot were also found to be higher in Pisco than in Ishinca.

The incidence of alpine turf removal (for tourist lodge floor and wall construction), shrub harvesting (for tourist lodge fuelwood) and seasonal burning for grass improvement was negligible in both valleys. These were found to be serious problems in the Everest region and were therefore investigated under the assumption that the same could hold true for the HNP alpine zone. The comparative lack of surficial disturbance related to shrub extraction is most likely related to the fact that bottled gas is brought in and used by nearly all tourist groups; alpine turf is not used for construction purposes (at least to the extent found in the Everest alpine region); and the practice of annually burning the pastures began to diminish some time ago (Tohan 2000, French 2003). The information was nevertheless useful in clarifying that while the overall impact was the same in both cases (that is, a highly disturbed alpine ecosystem), place-specific factors were the likely root causes of other forms of disturbance found.

These findings suggest that the overall condition of the alpine/high grassland ecosystems was worse than expected. The key disturbance factors contributing to groundcover change appear to be: (a) overgrazing; exacerbated by (b) unregulated adventure tourism, which has grown steadily since the 1960s. As mentioned previously, the dynamics of high altitude energy supply in the HNP differ substantially from conditions found in the Everest region, since propane or camp stoves are widely used in the HNP for cooking. Likewise, camp fires are rare and impacts on the local *Polylepis* forests appear to be minimal.[2] Similarities between the Everest and HNP regions, however, include degraded hillslope conditions related to overgrazing and the increased use of tourist-related pack

---

2 However, French (2000) reports that the refugios burn Polylepis on occasion, although the vast majority of wood observed by this study was Eucalyptus species. Author/climber Rick Ridgeway reports that during the 1960s, climbing expeditions burned large amounts of Polylepis in the basecamp regions (Ridgeway 2006, pers. comm.), a practice prohibited with the establishment of the national park in 1977.

animals (primarily donkeys); poor sanitary practices and conditions linked to the absence of toilet facilities; improper garbage disposal practices in both lower and upper basecamp regions; and a lack of awareness among all stakeholders (local communities, climbers, park officials) concerning the fragility, as well as economic importance, of the high alpine ecosystem.

Preventative and restorative solutions to these problems will revolve around a four-phase process that involves:

1. Understanding the site- or region-specific problems through applied, field-based research.
2. Reducing the threats.
3. Achieving the desired ecosystem and social/behavioural response.
4. Ensuring that long-term conservation mechanisms (for example memoranda of understanding, easements, official protection) are in place.[3]

The need for reliable biophysical and social data in the diverse and challenging mountain environment, and for the fullest understanding of problems prior to the prescription of solutions, has been covered in detail elsewhere (Ives and Messerli 1989, Ives 2004, Byers 2005). The second factor (above), namely the reduction of threats while achieving the desired results, will depend largely on the provision of viable alternatives to current grazing, landuse, sanitation and garbage practices, concurrent with educational campaigns designed to increase awareness and to change behaviour. This is a difficult task, but progress has been made on several fronts since the field work was conducted. Grazing pressures, for example, have been significantly reduced in the Ishinca alpine zone thanks to a TMI pasture improvement programme in the lower altitude village of Collón. Starting in 1998, five hectares of grazing land were enclosed, protected and improved near Collón, with the objectives of improving grazing practices, livestock health, productivity and income generation, and building local capacity to develop and manage better grazing systems. Eight years later, in 2006, the number of hectares protected by local communities has increased to 35, this land being used for both community cattle grazing as well as renting to *arrieros* for the grazing of their donkeys, llamas and horses used in the tourist trade. In 2007, an additional 15 hectares were protected in collaboration with local communities and with financial support from the Alpine Conservation Partnership. Livestock populations have been reduced significantly within the Ishinca Valley as a result of these interventions, and the high pastures and alpine zones of Ishinca are now allowed to rest during the rainy

---

3 I am indebted to Mr Rodney Bartgis, Director of The Nature Conservancy/West Virginia Chapter, for his suggestion of this monitoring model during the implementation of our joint project between 1999–2002: 'The Blister Swamp Conservation and Restoration Project, Pocahontas County, West Virginia'.

season when all livestock are grazed in the lower elevation exclosures.[4] Only the *chucaros* (feral cattle) remain within the core zone of the park. Demonstration exclosures within the alpine zone, however, could still assist in the restoration of alpine hillslopes, especially in the Pisco Valley, with the added benefit of relaying strong conservation and educational messages to local people as well as visitors.

Another model in need of further testing within the context of the alpine zone involves the use of 'conservation contracts'. For example, in the Ackya region due east of HNP, TMI assists community members to improve the quality of their pastures, and thus the quality and volume of milk production, in exchange for their active involvement in the restoration of biodiversity-rich *Polylepis* forests in the upper valley. In April 2006, more than 5,000 *Polylepis* seedlings had been grown and planted (this had risen to 10,000 seedlings planted per year in 2007 and 2008), and the level of participation and satisfaction among local *campesinos* appeared to be especially high. Experiences will differ community by community, but both projects are excellent examples of incorporating the actual linkages between livelihood improvements and on-the-ground conservation impacts within the project's overall design, effectively linking improved economics with improved conservation, and thus the long-term maintenance of various ecosystem services (such as water, hillslope stability, the cooling effect of continuous vegetation and forest canopies).

Basecamp and trekking camp sanitation is a problem that seems to grow worse with each passing year throughout the mountain world (Hillary, P. 2007, pers. comm.) and solutions remain evasive in spite of considerable efforts devoted to the development of simple, cost-effective backcountry toilets and high altitude sanitation strategies (see McConnell 1996, Lachapelle et al. 1997, French 2006a, Byers 2007f). The American Alpine Club, however, recently supported a pilot project to install a composting toilet in the Ishinca basecamp region in collaboration with the communities of Pashpa and Collón (French 2006b). In 2007, it was observed that the toilet was being utilized frequently by climbers in the region, but that its maintenance was problematic due to the lack of a full-time project overseer. Nevertheless, the initiative represents progress, especially when compared with the concrete and stone pit toilets installed throughout the HNP that become filled, dysfunctional and abandoned after a short period of time (French 2003, Byers 2007f). Similar experiences with high altitude outhouses are common throughout the developing mountain world, and I argue for accelerated research and continued trials that can develop safe and effective waste disposal mechanisms tailored to the high mountain environment.

4 During my August 2007 visit to the Ishinca Valley, I noted a much more continuous groundcover of grass and herbaceous species upon the alpine hillslopes, although the 2002 transects would have to be re-sampled to verify this impression.

Conclusions

Alpine environments worldwide are being heavily impacted upon, especially in regions experiencing high volumes of adventure tourism. Whilst oceans, deserts, rainforests and other ecosystems have strong(er) followings and conservation programmes, alpine ecosystems remain comparatively neglected in spite of their importance for global water supply, local economies and tourism. More interdisciplinary and applied research is essential for the better understanding of differing threats to alpine integrity, remedial interventions, capacity building mechanisms for local user groups and high altitude sanitation and waste management methods, and more attention needs to be focused on the large-scale conservation and restoration of many alpine ecosystems. Increasing the awareness of international alpine clubs, travel agencies and trekking companies for these problems will represent an essential first step in reducing user impacts. Building global awareness for the critically important ecosystem services that alpine and other mountain ecosystems provide, and the need for their better conservation and stewardship, also needs to be strengthened among conservationists, donor agencies and local communities alike.

Reducing threats, and achieving a positive ecosystem response, will depend largely on the provision to local people of viable alternatives to current grazing and landuse practices. The most promising 'long-term conservation mechanisms' required to protect alpine resources will also involve the building of local stakeholder capacity in project design, project management, financial management, fundraising, project monitoring and reporting. Education and awareness-building among all stakeholders (climbers, trekkers, indigenous people, governments and clubs) will be key to changing decades of detrimental trends in behaviour. The Alpine Conservation Partnership project represents the first global initiative to actively protect and restore alpine ecosystems while building the management capacities of the people who live there, with promising beginnings in the Himalayan and Andes mountains. Much work remains to be done, however, if the project's objectives are to be achieved in the hundreds of heavily impacted alpine ecosystems that remain throughout the high altitude mountain world.

References

Alirol, P. (1979) *Transhumant Animal Husbandry Systems in the Kalinchowk Region (Central Nepal): A Comprehensive Study of Animal Husbandry on the Southern Slopes of the Himalayas*. Weiss Association for Technical Assistance, Berne.

American Alpine Club (AAC) (2004) Khumbu reforestation project launched. *The American Alpine News*, Winter, 2004.

Ang, A. (2008) China plans Mount Everest cleanup in 2009. *Associated Press*, 23 June.

Barry, R.G. and Chorley, R.J. (2003) *Atmosphere, Weather and Climate* (8th edition). Routledge, London.

Bartle, J. (2004) Significant restrictions ahead for Huascarán National Park? *The American Alpine News*, Winter, 2004.

Bishop, N. (1989) From Zomo to Yak: change in a Sherpa village. *Human Ecology* 17, 177-204.

Bjönness, I.M. (1980) Animal husbandry and grazing, a conservation and management problem in Sagarmatha (Mt. Everest) National Park, Nepal. *Norsk Geografisk Tidsskr*, 34, 59–76.

Boukreev, A. and DeWalt, G.W. (1997) *The Climb: Tragic Ambitions on Everest*. St. Martin's, New York, NY.

Brako, L. and Zarucchi, J.L. (1993) *Catalogue of the Flowering Plants and Gymnosperms of Peru*. Monographs in Systematic Botany from the Missouri Botanical Garden, Vol. 45.

Brower, B. (1991) *Sherpa of Khumbu: People, Livestock, and Landscape*. Oxford University Press, Oxford.

Bury, J. (2007) *Collaborative Research: Glacier Recession and Livelihood Vulnerability in the Peruvian Andes*. Draft Report. Department of Geography, University of California, Santa Cruz, CA.

Byers, A.C. (1987a) *A Geoecological Study of Landscape Change and Man-\ accelerated Soil Loss: The Case of the Sagarmatha (Mt. Everest) National Park, Khumbu, Nepal*. Unpublished PhD thesis. Ann Arbor, University Microfilms International.

Byers, A.C. (1987b) Landscape change and man-accelerated soil loss: the case of Sagarmatha (Mt. Everest) National Park, Nepal. *Mountain Research and Development*, 7, 209–216.

Byers, A.C. (1996) *Repeat Photography of Mt. Everest National Park: Final Report*. National Geographic Society, Washington, DC.

Byers, A.C. (1997) Landscape change in the Sagarmatha (Mt. Everest) National Park, Nepal. *Himalayan Research Bulletin*, 17, 2. Special Issue: SoluKhumbu and the Sherpa.

Byers, A.C. (2000) Contemporary landscape change in the Huascaran National Park and Buffer Zone, Cordillera Blanca, Peru. *Mountain Research and Development*, 20, 52–63.

Byers, A.C. (2002) *New Roles for the Ecotourism Industry in the Protection of Mountain Environments*. Keynote presentation delivered at the International Ecotourism Society Summit, Quebec City, Canada, May 2002.

Byers, A.C. (2003) Landscape change in Sagarmatha National Park. In A.R. Sherpa and S. Höivik (eds) *Triumph on Everest: A Tribute from the Sherpas of Nepal*. Mandala Book Point, Kathmandu, pp. 91–100.

Byers, A.C. (2005) Contemporary human impacts on alpine landscapes in the Sagarmatha (Mt. Everest) National Park, Khumbu, Nepal. *Annals of the Association of American Geographers*, 95, 112–140.

Byers, A.C. (2007a) *A Preliminary Survey of Contemporary Human and Cattle Impacts on Alpine and Subalpine Ecosystems of the Hinku Khola and Mera Peak Regions, Makalu-Barun National Park, Nepal, 12–22 May, 2007: Problems and Prospective Solutions*. Progress Report to the Argosy Foundation. The Mountain Institute, Elkins, WV.

Byers, A.C. (2007b) The Alpine Conservation Partnership: a global initiative to protect and restore alpine ecosystems. *The American Alpine Journal*, 81, 121–123.

Byers, A.C. (2007c) *A Preliminary Assessment of Trekker Impacts on Mt. Kilimanjaro, Tanzania*. The Mountain Institute, Elkins, WV.

Byers, A.C. (2007d) An assessment of contemporary glacier fluctuation in Nepal's Khumbu Himal using repeat photography. *Himalayan Journal of Sciences*, 4, 21–26.

Byers, A.C. (2007e) Impacts of climate change in the mountains. In G. Braasch (ed.) *Earth Under Fire*. University of California Press, Berkeley.

Byers, A.C. (2007f) *Trip Report: Huascarán National Park, Ancash Department, Peru, 3–31 August, 2007*. The Mountain Institute, Elkins, WV.

Byers, A. and Banskota, K. (1992) Environmental impacts of backcountry impacts on three sides of Everest. In J. Thorsell and J. Sawyer (eds) *World Heritage Twenty Years Later*. IUCN, Gland, pp. 105–122.

Carpenter (1993) *An Overview of the Biological Ecology of Makalu-Barun National Park and Conservation Area, Sankhuwasaba District, Nepal*. Wildlands Studies Program. San Francisco State University, San Francisco.

Chepesiuk, R. (2003) Party is over but degradation of Everest persists. *The Daily Star*, 20 June.

Coburn, B. (1997) *Everest: Mountain Without Mercy*. National Geographic Society, Washington, DC.

Coppock, R. (1978) The influence of Himalayan tourism on Sherpa culture and habitat. *Zeitschrift fur Kulturaustausch*, 28, 61–68.

Cox, J.H. (1999) *An Assessment of Habitats and Human Interaction in the Hinkhu, Hongu, Kasuwa and Barun Kholas of Makalu-Barun National Park and Conservation Area*. Makalu Barun Conservation Project, Kathmandu.

Deegan, P. (2003) Appetite for destruction. *Geographical*, March 2003, 32–36.

Dourjeanni, P. (2006) Program Officer, The Mountain Institute. Personal communication, Huaraz, Peru, August, 2006.

Fisher, J.F. (1990) *Sherpas: Reflections on Change in Himalayan Nepal*. University of California Press, Berkeley, CA.

Fjeldså, J. and Kessler, M. (1996) *Conserving the Biological Diversity of Polylepis Woodlands of the Highlands of Peru and Bolivia: A Contribution to Sustainable Natural Resource Management in the Andes*. Nordic Foundation for Development and Ecology (NORDECO), Copenhagen.

Foro Ecológico del Perú (2006) *Petition to the World Heritage Committee requesting inclusion of the Huascaran National Park in the list of World*

Heritage in Danger as a result of Climate Change. Foro Ecológico del Perú, Lima, Peru.

French, A. (2003) *A Preliminary Analysis of Adventure Tourism Impacts in Huascarán National Park*. Unpublished Report. Department of Environmental Studies, University of Montana, MT.

French, A. (2006a) *Agua es vida—Water is Life: Promoting Community and Watershed Health in Peru's Cordillera Blanca*. Proposal to the American Alpine Club, Zack Martin Breaking Barriers Grant Program.

French, A. (2006b) *The Ishinca Peace Toilet*. The American Alpine Club, Golden, CO.

Fürer-Haimendorf, C. von (1984) *The Sherpas Transformed*. Sterling Publishers Private Limited, New Delhi.

Haffner, W. (1972) Khumbu Himalaya: Landschaftsokologische untersuchungen in den Gochtalern des Mt. Everest-Gebietes. In C. Troll (ed.) *Geoecology of the High-Mountain Regions of Eurasia*. Franz Steiner Verlag GMBH, Wiesbaden, pp. 244–263.

Hemmleb, J., Johnson, L., and Simonson, E. (1999) *Ghosts of Everest: The Search for Mallory and Irvine*. The Mountaineer Books, Seattle.

Hillary, P. (2007) Mountaineer, explorer, and motivational speaker. Personal communication, Melbourne, Australia, March 2007.

Instituto Nacional de Estadistica e Informatica (INEI) (2002) *Huascarán National Park Buffer Zone*. Available at: http://www.inei.gob.pe/.

Instituto Nacional de Recursos Naturales (INRENA) (1996) *Plan de Uso Turistico y Recreativo del Parque Nacional Huasraran* (Vol. 1). Instituto de Montaña, Lima.

Instituto Nacional de Recursos Naturales (INRENA) (2003) *Parque Nacional Huascarán Plan Maestro 2003–2007*. INRENA, Lima.

Ives, J.D. (2004) *Himalayan Perceptions: Environmental Change and the Well-Being of Mountain Peoples*. Routledge, London.

Ives, J.D. and Messerli, B. (1989) *The Himalayan Dilemma*. Routledge, London.

Klatzel, F. and Byers, A.C. (2001) *Celebrating the People and Landscapes of Mount Everest and the Khumbu Valley, Nepal*. Seventeen-panel photographic exhibit on permanent display in the Sagarmatha National Park Museum. See: http://www.mountain.org/work/resedu/khumbu.htm.

Körner, C. (1999) *Alpine Plant Life: Functional Plant Ecology of High Mountain Ecosystems*. Springer-Verlag, Berlin.

Krakauer, J. (1997) *Into Thin Air*. Anchor Books/Doubleday, New York.

Lachapelle, P. (1998) Managing sanitation in protected areas: problems and challenges in Sagarmatha (Mt. Everest) National Park, Nepal. *Himalayan Research Bulletin*, 18, 53–57.

Lachapelle, P., Land, B.L., and Clark, J.C. (1997) The problem of human waste disposal in national parks: a solar energy experiment in Yosemite National Park, California, CA. *Mountain Research and Development*, 17, 177–180.

Lipton, J. (2002) *Land use and Land cover Change in and around Huascarán National Park, Ancash, Peru using Time-series Landsat ETM + Satellite Imagery.* Unpublished Report. Department of Geography, University of Texas Austin, TX.

Management Bureau of the Qomolangma Nature Preserve (1989) *Kharta, Pazhuo, and Rongbuk Base Camp Area Survey Report*. Qomolangma Nature Preserve, Shigatze.

Mani, A. (1981) The climate of the Himalaya. In J.S. Lall (ed.) *The Himalaya: Aspects of Change*. Oxford University Press, Oxford, pp. 3–15.

March, K. (1977) Of people and nak: the management and meaning of high altitude herding among contemporary Solu Sherpa. *Contributions to Nepalese Studies*, 4, 83–97.

Margoluis, R. and Salafsky, N. (1998) *Measures of Success: Designing, Managing, and Monitoring Conservation and Development Projects*. Island Press, Washington, DC.

McConnell, B. (1996) *Gentle Expeditions: A Guide to Ethical Mountain Adventures*. American Alpine Club Press, Golden, CO.

Meihe, G. (1987) An annotated list of vascular plants collected in the valleys south of Everest. *Bulletin of the British Museum (Natural History) (Botany Series)*, 16, 225–268.

Meihe, G. (1989) Vegetation patterns on Mount Everest as influenced by monsoon and föhn. *Vegetatio*, 79, 21–23.

National Geographic Society (2003) *Everest 50*. Map of the Everest region included in the May, 2003 issue of National Geographic Magazine in celebration of the 50th anniversary of the climbing of Mt. Everest.

Naylor, R. (1970) *Colombo Plan Assignment in Nepal*. Draft Report. New Zealand Forest Service, Rotorua.

Nepal, S.K. (2000) Tourism in protected areas: the Nepalese Himalaya. *Annals of Tourism Research*, 27, 661–681.

Nepal, S.K. (2001) Examining tourism impacts from an interdisciplinary perspective: the Himalayan case study. *Asiatische Studien Études Asiatiques*, LV–3–2001, 777–804.

Nepal, S.K., Kohler, T., and Banzhaf, B.R. (2002) *Great Himalaya: Tourism and the Dynamics of Change in Nepal*. Swiss Foundation for Alpine Research, Berne.

Ortner, S.M. (1999) *Life and Death on Mt. Everest: Sherpas and Himalayan Mountaineering*. Oxford University Press, Oxford.

Parkswatch (2005) *Park Profile: Huascarán National Park, Peru*. Available at: http://www.parkswatch.org/main.php.

Price, M.F. and Barry, R.G. (1997) Climate Change. In B. Messerli and J.D. Ives (eds) *Mountains of the World: A Global Priority*. The Parthenon Publishing Group, New York and London, pp. 409–445.

Recharte, J. (2006) Director, Andes Program, The Mountain Institute. Personal communication, Huaraz, Peru, August 2006.

Rhoades, R., Rios, X.Z., and Ochoa, J.A. (2006) Mama Cotacachi: history, local perceptions, and social impacts of climate change and glacier retreat in the Ecuadorian Andes. In B. Orlove (ed.) *Darkening Peaks*. University of California Press, CA, pp. 216–228.

Ridgeway, R. (2006) Marketing Director, Patagonia, Inc. Personal communication, Washington, DC, October 2006.

Sagarmatha National Park (2008) *Tourist Numbers 1998–2007*. Display, Jorsalle entrance station, Solu Khumbu, Nepal.

Sherpa, A.R. (2003) *Alpine Conservation Restoration Program in Khumbu Region of Sagarmatha National Park and Buffer Zone*. The Mountain Institute, Kathmandu.

Sherpa, A.R. (2004) *Community Planning Workshop: Alpine Conservation and Restoration Planning, Dingboche, May 2004*. The Mountain Institute, Kathmandu.

Sherpa, A.R. (2008) *Alpine Area Conservation: A Community Approach*. ECS, May 2008. Kathmandu, Nepal.

Sherpa, L.N. (2001) Director, Qomolangma Conservation Project. Personal communication, Kathmandu, Nepal, March 2001.

Sherpa, N.W. (2001) Trekking sirdar. Personal communication, Sagarmatha National Park, Nepal, September 2001.

Smith, D.N. (1988) *Flora and Vegetation of the Huascarán National Park, Ancash, Peru, with Preliminary Taxonomic Studies for a Manual of the Flora*. Unpublished PhD thesis, Iowa State University, IA.

Stevens, S.F. (1993) *Claiming the High Ground: Sherpas, Subsistence, and Environmental Change in the Highest Himalaya*. University of California Press, Berkeley, CA.

Stevens, S.F. (2003) Tourism and deforestation in the Mt Everest region of Nepal. *The Geographical Journal*, 169, 255–277.

Taylor-Ide, D. (2002) Director, Future Generations, Franklin, WV. Personal communication.

Tohan, A.K. (2000) *Polylepis Forests of the Andes: A Socio-Environmental Case Study from the Huascarán Biosphere Reserve, Ancash, Peru*. Unpublished Masters thesis. Department of Geography, University of Wisconsin-Madison, WI.

Tovar, O. and Oscanoa, L. (2002) *Guía para la identification de pastos naturals alto andinos de mayor importancia ganadera*. Instituto de Montaña, Huaraz.

Union Internationale des Associations D'Alpinisme (UIAA) (2003) *Mountaineers fund new Everest conservation project: American Alpine Club and The Mountain Institute join forces*. UIAA Online, 2 June 2003.

Zimmerman, M., Bichsel, M. and Kienholz, H. (1986) Mountain hazards mapping in the Khumbu Himal, Nepal, with prototype map, scale 1:50,000. *Mountain Research and Development*, 6, 29–40.

Chapter 5
Tourism, Local Community and Natural Resources: Tourism Impact Assessment and Tourism Management Analysis in the Jiuzhaigou Biosphere Reserve, China

Wenjun Li

Introduction

Tourism development may result in a range of negative impacts on the environment, emanating from unsustainable resource use and mismanagement (Lukashina et al. 1996, Collins 1998, Ross and Wall 1999, Kousis 2000, Nepal 2000, Van der Duim and Caalders 2002, Laiolo 2003, Mbaiwa 2003, Kuvan 2005). Nevertheless, the development of tourism may also bring positive influences by offering an alternative livelihood to preceding unsustainable natural resource utilization such as farming, hunting and mining. Tourism has the potential to contribute to both conservation and development (Ross and Wall 1999) and it involves the creation of positive synergetic relationships between tourism, biodiversity and local people through the application of appropriate management strategies. Examining tourism development and its relationship to natural capital, Collins (1998) lists a substantial body of research relating to the notion of sustainable tourism development (see references therein). Burger (2000) explored three examples of the interactions between humans and ecosystem integrity at the landscape scale, and concluded that tourism/recreation can theoretically play a major role in preserving biodiversity. Nevertheless, there are few practical examples that demonstrate the positive environmental impacts caused by tourism. In other words, it appears that sustainable tourism theory has rarely been put into practice successfully (Ross and Wall 1999).

Given that tourism presents a 'double-edged sword' for natural resource conservation, clarifying its positive and negative effects would prove helpful in gaining a more comprehensive understanding of the relationship between the industry, local communities and natural resources, and it would also provide a valuable guide in natural resource management.

In this chapter, taking Jiuzhaigou Biosphere Reserve (JBR) as a case study area and employing on-site questionnaire surveys, field interviews, and remotely sensed satellite image interpretation, a positive case study is presented whereby

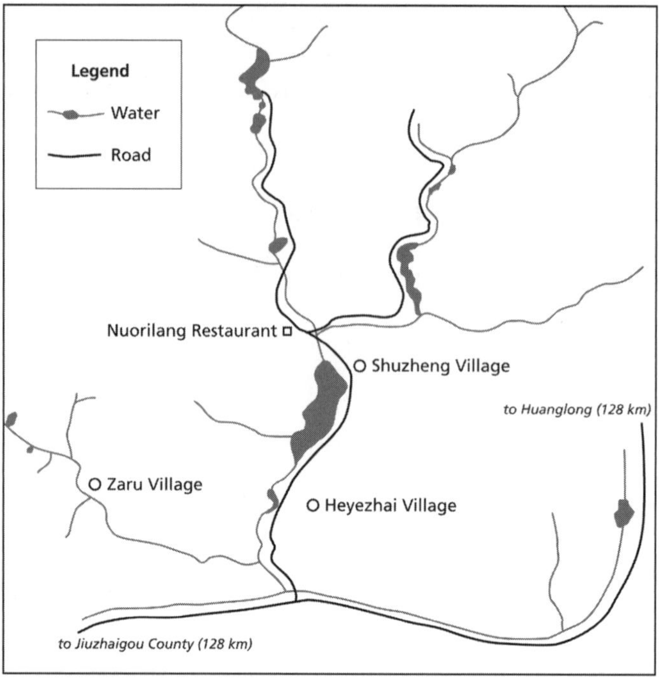

Figure 5.1 Distribution of administrative (and case study) villages within Jiuzahigou Biosphere Reserve

the environment is not widely degraded as tourism develops and the indigenous community and local region benefit from the industry. The development and operation of a joint-stock company to manage tourism, established between local community households and reserve administrators, is analysed in order to identify the foundation to this particular story.

Field Site and Methods

Case Study Area

The JBR is located within China's Sichuan Province (between 103°46′E and 104°05′E and 32°55′N and 33°16′N). The reserve adopts the form of a Y-shaped valley, with a length of 40 km and a total area of 720 km^2. Its elevation ranges from 2,000 m to 4,764 m. Resources for tourism are abundant in Jiuzhaigou County and the reserve consequently attracts many visitors each year. Numerous snow-topped peaks, including modern glaciers, are distributed along the valley. The geology of the reserve is comprised of dissolvable carbonate layers that are famous for their travertine phenomena. Denuded sediments adhere to the beds of lakes and rivers, resulting in colourful vistas, and scenic spots include imposing emerald lakes, waterfalls and forests. The area is rich in floral and faunal biological diversity, containing, for example, over 4,000 plant species and 123 vertebrate species.

In addition to the natural scenic beauty and biodiversity, traditional culture is also an attraction for tourists. Three administrative villages comprise the local community of the JBR, namely Heyezhai, Shuzheng and Zaru (Figure 5.1). Of the total population of 1,097, distributed in 269 households within the reserve in 2004, Tibetans account for over 80 per cent.

Questionnaire Surveys and Interviews

A series of questionnaire surveys were conducted with the residents of the JBR during the years 2000 to 2005. The first survey, conducted in July 2000, was designed to evaluate economic, social and environmental aspects of tourism development. The economic indicators included the number of people involved in and receiving benefits from tourism, the nature of their involvement, and how much of their income was derived from tourism. The social assessment was conducted by asking residents to rank their perceptions about interference from tourism development (including noise, personal safety, cultural change and commodity prices). Respondents were asked to check one answer in the questionnaire from a choice of four possible statements: 1) No interference; 2) Slight interference; 3) Strong interference; and 4) Extreme interference. The environmental assessment was conducted by ranking the perceptions of respondents in terms of environmental change since tourism developed (including impacts on wildlife, landscape, water and atmosphere). There were four choices for selection in the questionnaire: 1) No degradation; 2) Slight degradation; 3) Strong degradation; and 4) Extreme degradation. As the local populace has inhabited the region for generations, it is likely that they are sensitive to environmental change.

The questionnaire was designed and executed in Chinese because most of the Tibetan residents in the JBR can understand and speak this language. The surveys were conducted face to face, and all questionnaires were filled out by the researchers as questions were answered by the local residents. This method was adopted as some of the residents were illiterate and many felt nervous about completing the questionnaire directly. The surveys were conducted in a conversational style, so the residents were relaxed and willing to speak freely about pertinent issues. Most of the questionnaires were answered by the head of household, whether the wife or husband. Surveyed households were sampled randomly at a 20 per cent intensity in the three villages.

In the summers of 2001 and 2002, open-ended interviews were conducted with senior managers of the JBR to understand the policies implemented for encouraging local community involvement in tourism and the measures taken for sustainable tourism management.

After the above surveys, it was clear that the managers of the JBR were using an innovative joint-stock system to ensure that benefit accrued to the local community.

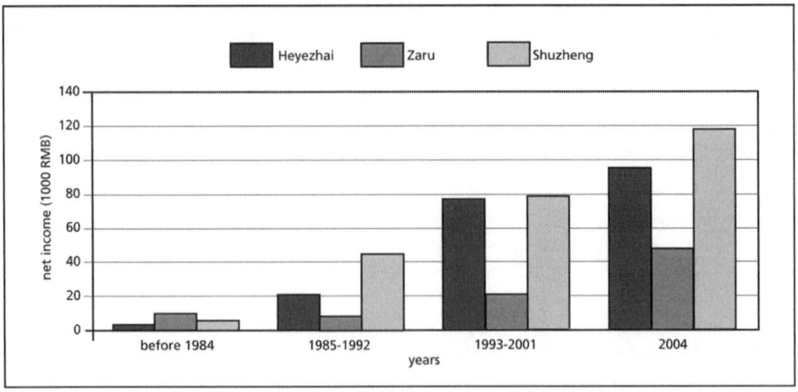

Figure 5.2 Comparison of annual net income per household among the three case study villages in Jiuzhaigou Biosphere Reserve

Source: Author household surveys.

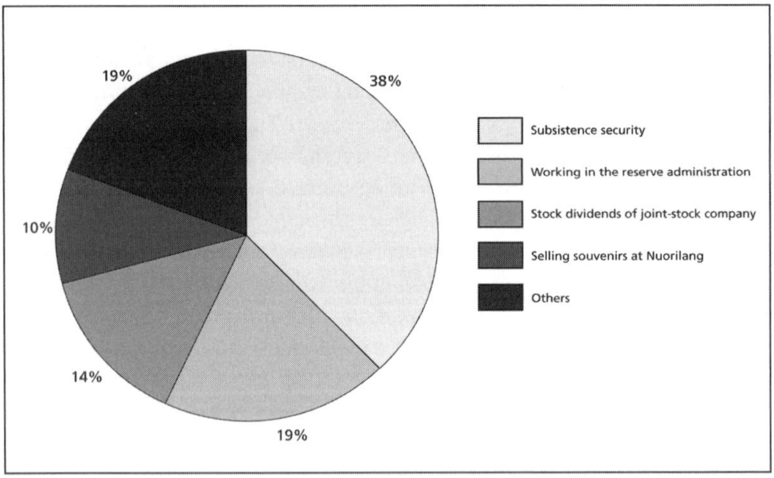

Figure 5.3 Proportions of different incomes in the Jiuzhaigou Biosphere Reserve community in 2004

Source: Author household surveys.

As a consequence, a second questionnaire survey and further interviews were designed and conducted from November 2004 to August 2005. These collected information related to the mechanism established to organize tourism between the reserve managers and the local residents. The questionnaire survey was conducted at households within the reserve, and it was designed to elucidate the history of local resident participation in tourism, the origin and development of the joint-stock system, its effectiveness, its impact on the income of local households, and any pervading problems. As in the year 2000, the same sample intensity of 20 per cent of households was adopted. Open-ended interviews were simultaneously conducted with the JBR managers as well as with the directors of the joint-stock company. Finally, selected officials at different levels of government were interviewed to collect their opinions and attitudes on the joint-stock system, including ZhangZha Township governments, Jiuzhaigou County governments, and Aba Prefecture governments.

In addition to the on-site surveys, the following secondary data were collected from the JBR management and local government: tourism revenue and contribution to the income of local people, economic structural change, employment structural change, and other relevant policy documents.

Remote Sensing Image Interpretation

Remotely sensed satellite images were analysed to quantify environmental change over space and time in Jiuzhaigou County. In order to identify whether the pressures from local residents who live directly on natural resources have been transferred from inside to outside of the JBR, three years of Landsat Thematic Mapper (TM) images were selected for spatio-temporal analysis: July 1987, September 1997 and September 2003. This time series was used to identify land cover changes in the county. The image from 1987 indicated the land cover situation at the beginning of tourism development; that of 1997 marked the transition point from which visitor number increased dramatically and prevailing land cover was determined from the 2003 image.

Considering the geological, physiognomic and biological characteristics of Jiuzhaigou County, TM bands 5, 4, and 3 were selected to compose a false colour image and this was used in the interpretation. First, a non-supervised classification was undertaken to obtain an initial interpretation image. This included seven classes covering forest, shrub land, grassland, bare land, water, farmland and residential land. An on-site field validation was conducted in July 2003 and a total of 121 sites covering the above seven classes were located using a global positioning system. Using these on-site field training samples, a maximum likelihood method of supervised classification was applied to further improve the accuracy of the image and resultant interpretation.

Results

The Economic Contribution of Tourism to Local Residents

The tourism history of the JBR can be divided into four periods according to different stages of development. The years preceding 1984 can be defined as the pre-tourism period, during which there were few tourists and no substantial tourism

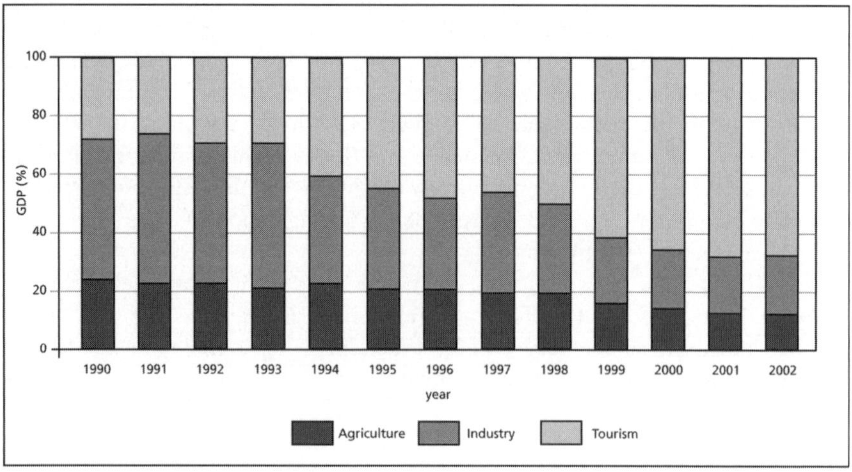

Figure 5.4 Changes in GDP in different economic sectors from 1990 to 2002 in Jiuzhaigou County

Source: Redrawn from Fig. 5, Li et al. (2006) Tourism's impacts on natural resources: a positive case from China. *Environmental Management*, 38, 577.

services provided in the natural reserve. The period from 1985 to 1992 represented the initial stage of tourism development. During this period, the number of tourists increased to 139,000 per year and many households in Shuzheng and Heyezhai villages operated individually-owned family hotels that became their main source of income. The years between 1993 and 2001 marked a period of rapid tourism development, during which the number of tourists increased to 1.2 million per year and a joint-stock company was established between reserve managers and local households to operate family hotels and restaurants. Finally, the time between 2002 and 2004 represented a period of maturity in tourism development, with the number of tourists stabilizing at around 1.8 million per year and the joint-stock company operating just one concern, Nuorilang Restaurant, after all family-run hotels were closed in 2001.

Before tourism was developed, the local community lived a self-sufficient but impoverished life, dominated by agriculture and grazing. The annual per capita income was around US$23 in 1978. The community had limited sources of income, and it depended strongly on direct exploitation of natural resources via farming, woodcutting, hunting and so on. After tourism was developed in 1984, local income increased dramatically (Figure 5.2). Compared with 1978, the per capita income in 1988 was increased by 188 per cent (after 10 years), and by 477 per cent in 1999 (after 20 years). This great increase in community income was closely linked with the vigorous development of tourism. At present, most community incomes are comprised of monies from tourism development, derived both directly and indirectly. As shown in Figure 5.3, the first and most important income is 'subsistence security', a share of money allocated from entrance fees

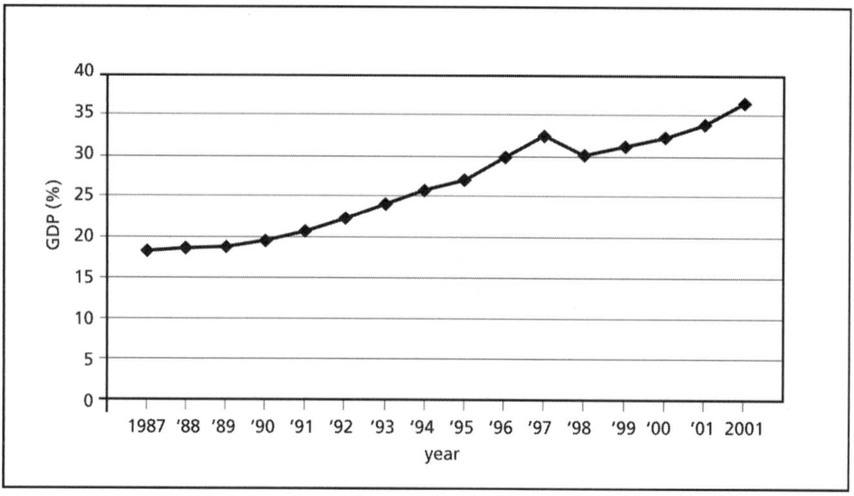

Figure 5.5 Proportion of the population living on non-agricultural activities

Source: Redrawn from Fig. 6, Li et al. (2006) Tourism's impacts on natural resources: a positive case from China. *Environmental Management*, 38, 577.

to the reserve to assure local subsistence. This will be described in detail later in the chapter. Subsistence security accounts for 38 per cent of the total incomes of local residents. The second most important income source is the salary paid by JBR managers to workers in the reserve (19 per cent), followed by earning from dividends of the joint-stock company, which account for a further 14 per cent of the total. Income from selling souvenirs and renting costumes for photographs comprises 10 per cent of the incomes of local residents.

The questionnaire survey results showed that 99 per cent of the households selected 'No interference' in terms of perceived impacts of tourism on the local socio-economic and physical environments. This result indicated that the local residents had a high tolerance for the impacts of tourism, possibly reflecting an active involvement in the industry and the receipt of expected benefits.

The Contribution of Tourism to the Regional Economy

Rapid tourism development in the JBR has dramatically driven the economic development of Jiuzhaigou County and caused notable changes in both the economic sector structure and employment structure. The Gross Domestic Product (GDP) from the tourism service sector increased from 28 per cent in 1990 to 68 per cent in 2002 (Figure 5.4). Over the same period, the agricultural GDP decreased from 28 per cent to 11 per cent. The local labour employed in the tourism service sector doubled from 15 per cent in 1982 to 32 per cent in 2002. With the local economic structure changes, the percentage of non-agricultural population increased from

18 per cent in 1987 to 37 per cent in 2002 (Figure 5.5). As livelihoods shift from the agricultural to the tourism sector, pre-existing pressures on natural resource conservation from within and from peripheral areas of the JBR might possibly be mitigated. This is examined in the following environmental assessment.

Natural Environmental Impacts

From the questionnaire survey with local residents, 90 per cent of interviewees selected the option of 'No degradation' in terms of perceived environmental changes within the JBR. Some respondents even indicated that the natural landscape had improved in their view since the 1990s, coinciding with villagers ceasing their agricultural activities due to receipt of viable income from tourism. As the environment has shown signs of recovery, the number of wild animals seems to have increased. A forest ranger, who worked in Jiuzhaigou County for a state lumber sector in the 1970s and who has been working in a protection role since reserve status was approved in 1979, commented via the questionnaire survey that 'the chance we meet wildlife when we patrol in the core area in recent years is obviously more than in the 1970s when we conducted clearing cutting there'. This result coincides with the report of the Sichuan Provincial Academy of Forestry Sciences (SPAFS) (2004). This organization conducted a comprehensive scientific survey in the JBR between 2002 and 2003, and reported: 'Currently, the forest, shrub land and meadow are [following] the normal and natural dynamic evolution process'.

The results from the satellite image analysis for the years 1987, 1997 and 2003 are shown in Table 5.1. It can be seen from this table that forest area decreased by 8 per cent between 1987 and 1997, whereas it increased by 2 per cent thereafter. The area of shrub land increased continuously from 1987 to 2003, though its rate of recovery slowed after 1997; grassland increased in area by 22 per cent between 1987 and 1997 and remained stable thereafter. Generally, it appears that the vegetation (forest, shrub land and grassland) is experiencing a recovery from previous disturbance. There is evidence that the mid stages of vegetation succession have experienced the greatest increase in area and this may bode well for future successional development to forest. Additionally, the area of bare land decreased, although an increase was recorded initially from 1987 to 1997. This natural regeneration of vegetation must, however, be weighed against the increase in residential land. Residential land area first decreased between 1987 and 1997, and then increased dramatically due to the large-scale in-migration of indigenous people in 1992 from high elevations, where they traditionally lived, to lowland areas near popular scenic spots where family hotels could be operated with relative ease. When the indigenous population inhabited the highlands, their houses were widely dispersed. After migration to the lowland, to take part in the tourism industry, the local population was required by the JBR Administration (JBRA) to adopt a more nucleated pattern of settlement to ensure minimal influence on sightseeing. As a final comment on Table 5.1, the dramatic increase in water area might be explained

Table 5.1 Percentage changes in land cover from 1987 to 2003 in Jiuzhaigou County

	1987–1997	1997–2003
Forest	-8	+2
Shrubland	+131	+11
Grassland	+22	0
Water	+33	+229
Farm land	-26	-61
Residential land	-28	+119
Bare land	+5	-15

Source: Table 1, Li et al. (2006) Tourism's impacts on natural resources: a positive case from China. *Environmental Management*, 38, 578.

by varying annual rainfall. Because the acquisition dates of the three images are all during the peak rainfall period (May to October), the substantial spatial changes in water recorded in the images might result from the differences between the three annual rainfall totals. Unfortunately, the relevant regional rainfall data were not available to test this assumption.

Policies and Mechanisms Ensuring Indigenous Resident Benefits

Policies encouraging residents to participate in tourism From the interviews with JBR managers, it was noted that many policies and mechanisms had been put into practice to encourage local residents to participate in tourism. These measures enabled the local residents to derive enough economic benefit from tourism to abandon direct exploitation of natural resources for subsistence. One policy at the beginning of tourism development was to organize local residents to operate family hotels to host visitors. The first family hotel operator told us: 'It was the JBR administration and local governors who encouraged and helped me to borrow money from the bank, so I could be the first person to decorate my house and host visitors in 1984'. He also said excitedly: 'At the beginning, nobody in the village believed there would be visitors who liked to live in my house. Yes, they didn't believe I could make money from tourism, but after one year every household wanted to operate a family hotel because I suddenly became a rich man through the family hotel business'. For those residents who lived further away from scenic spots (and so could not make money through family hotels), the JBRA gave them priority to apply for tourism service jobs. The administration also paid every household 450 RMB/month (US$54) from the reserve entry fee for the purpose of protecting forests.

Origin and development of the joint-stock system The development of the joint-stock company can be divided into two periods; the family hotel period (1993 to 2001) and the Nuorilang Restaurant period (post-2001).

At the beginning of tourism development, in the early 1980s, the majority of households in Jiuzhaigou wanted to develop a family hotel after the first such hotel proved successful (Li 2006), and the number of beds increased rapidly without consideration of the total quantity of tourists. This lack of market management resulted in almost all the local hotel-owners standing daily at the reserve entry gates during the peak season, contending for tourists to accommodate in their family hotels, and sometimes adopting unreasonable price discounting. This situation was viewed negatively by the tourists and it impeded tourism development. Conscious of this situation, in 1990 ten family hotels in Shuzheng Village combined together and set up a joint company to consolidate their family hotels and home restaurant operations. The company unified service price and quality, thereby avoiding *in* the village the intense competition for tourists that still existed *amongst* the different villages. With this success in mind, in 1992 the JBRA organized and combined all the family hotels and household restaurants operating in the reserve with the restaurant operated by the JBRA itself to establish a Jiuzhaigou Tourism Joint-Stock Company (renamed as the Joint Operating Company, or JOC, in 1999). The family hotels were appraised as capital stock to join the company, and residents shared most of the stock (77 per cent). The remaining 23 per cent of the stock was held by the JBRA itself, the income being used for management of the nature reserve. In order to keep tourist numbers below carrying capacity, each family hotel was not allowed to exceed 45 beds. Excluding a small number of families that did not have enough capital to offer that number, almost all the family hotels did offer 45 beds. Instead of the previous intense competition, tourists who needed lodging were appointed to family hotels by the JBRA. Tourists paid accommodation fees directly to the JOC, instead of to the owners as had been done in the past. At the end of each year, accommodation fees were allocated by the JBRA to each family that operated a hotel.

With the growth of tourism in the valley, the number of family hotels increased greatly and this resulted in environmental pollution. In 1996, the Sichuan Environmental Monitoring Centre discovered that the water quality in the Jiuzhaigou natural reserve was lower than in 1986, and the river sections where water quality was worse were all near the family hotels and home restaurants (SPAFS 2004). In order to prevent any further environmental pollution, in 1997 the JBRA forbade local families to build any new hotels. Furthermore, on 30 April 2001, all the family hotels were closed as dictated by the administration. As the family hotel was the major source of income for the residents, the JBRA has paid subsistence security fees annually to residents living within the reserve since that date. This has amounted to approximately 7.2 thousand RMB per person annually.

Although family hotels were closed, the JOC continued to distribute revenue from the restaurants. The restaurants' benefits decreased dramatically, however, once tourists remained in the valley only during the daytime due to there being no accommodation after the closure of the family hotels. As a consequence, the JOC decided to keep only one restaurant, at the Nuorilang scenic site, open (instead of the original five), and the Nuorilang Restaurant was rebuilt and enlarged (Figure 5.1). Once again, the new Nuorilang Restaurant was established in joint-stock mode. The JBRA encouraged local residents to join this company by purchasing stock at a cost of 2,000 RMB per share, with each family allowed to own a maximum of 10 shares. The JBRA retained 51 per cent of the stock, whilst local residents obtained the remaining 49 per cent. Profits were distributed according to the percentage of shares owned by stockholders, which resulted in the JBRA receiving 51 per cent of the profits and individual stockholders together receiving 49 per cent. The JBRA then passed 28 per cent to the resident management office, the unit in charge of managing residents' issues, to build and improve infrastructure for the local community. Consequently, the final allocation of profits was 23 per cent to the JBRA and 77 per cent to the local communities.

Benefits of the joint-stock mechanism As described above, JOC shared the stocks with local residents and thus created a mechanism to ensure their economic benefit. As shown in Figure 5.2, the annual net income per household of Shuzheng and Heyezhai villages increased dramatically after tourism development, especially post-1992, when JOC was established to unify the operation. According to the questionnaire data, approximately 86 per cent of Shuzheng Village residents and 91 per cent of Heyezhai Village residents are currently living on tourism.

In addition to increased income, the joint-stock mechanism effectively reduced the income gap between villages. Zaru Village could not get involved in the tourism operation because it was located too far from the most attractive scenic spots (Figure 5.1). The main income resource of residents has remained with crop farming and their income has been much lower than the residents of Shuzheng and Heyezhai. In 2001, Zaru residents were encouraged to join stock in Nuorilang Restaurant under the coordination of the JBRA and, as a result, their overall income increased rapidly. Figure 5.2 shows that the income gap in 2004 was much smaller than the period from 1992 to 2001. Questionnaire responses in Zaru Village showed that an average household's profit from the JOC, together with gift selling, accounts for 30 per cent of their total income.

As described in the development history of the joint-stock mechanism, the JBRA unified the tourism operations in the natural reserve, eliminated the intense competition between households and maintained the Jiuzhaigou market image. Furthermore, under the current imperfect institutional arrangements for conservation (such as the lack of nature reserve law and the fact that current regulations are not strict and operational), the JBRA can put conservation ideas into practice through controlling the economy. For example, the reason why the administration could close the unsustainable family hotels within two years was due to it providing sufficient economic compensation to the owners. Also, about 38 per cent of household income is now derived from subsistence security fees that are decided upon and provided by the JBRA, and these are paid on the precondition that householders are mindful of conservation. Of course, this will also bring negative influences, related to equity issues in the future, but this issue is beyond scope of this chapter.

From the point of view of environmental conservation, the most important goal of the JBRA is protecting natural resources. Affiliated branch companies do not purely pursue profit maximization; sometimes they forego economic benefits to protect nature. In the design and construction of Nuorilang Restaurant, for example, the JBRA considered its potential impacts on the environment. As a result, sewage is collected in tanks and is then transported outside of the reserve. Such measures increased the cost of constructing the restaurant in the short term, but they benefit the environment in the long term.

Discussion and Conclusions

Forest cover was reduced in Jiuzhaigou County at the beginning of tourism development in the mid-1980s, before it increased after 1997 (Table 5.1). This result indicates that environmental decline from tourism was overcome in Jiuzhaigou County in a relatively short time, as evidenced through ecological restoration via vegetation succession. The area of shrub land and grassland increased, whilst farmland and bare land decreased. This improvement in the natural environment of Jiuzhaigou, however, has to be balanced against a notable increase in residential land after 1997.

Field survey results indicated that local residents in the JBR believed that the physical environment in the reserve was improving and they also felt that they were receiving satisfactory benefits from tourism. It is asserted that the main reason for the positive impacts of tourism in the JBR and in the wider Jiuzhaigou County is that local community residents participate fully in the tourism business and hence benefits from it are thereby ensured. Only when local communities are involved extensively in a tourism business can their original natural resource exploitation be substituted and the environment experience recovery.

From this case study, it can be seen that tourism does not necessitate negative environmental impacts. If indigenous residents benefit sufficiently from tourism, it is possible to use the industry as a way to balance natural resource conservation and economic development. In this case study, effective policy in the form of a joint-stock mechanism, which ensured local participation in and benefit from tourism, is the foundation supporting success.

Two additional purposes of the study were to determine whether the JBR model could be replicated in other protected areas and whether it could be sustained over the long term. To answer these questions, it must be understood why this mechanism was selected and how it was applied in the JBR.

In 1992 the government stopped all financial support to the JBR. This meant that the reserve had to ensure its survival and conduct conservation entirely through self-support. This resulted in reserve administrators drawing the natural resources in their care into the market economy. But why did they decide upon a joint-stock company as the operative mechanism? One reason was that in the early 1990s joint-stock companies were encouraged by central government and were widely used in China's state enterprises. They were believed to be an effective way to obtain more capital investment from non-government sectors and to lead to high economic efficiency. So the selection of a joint-stock mechanism in the JBR could not be isolated from this background of national economic policy.

By the end of the 1990s, many reserve managers selected to contract out tourism services to private companies. This ignited a heated debate on the adoption or rejection of market approaches in China (X.Z. Zhang 2000, 2002, X. Zhang 2005, 2006). When we asked the Director General why he did not opt to contract out the tourism business as did the management of Siguniang Mountain Scenic Site (located in the same province and close to the JBR), he replied 'We don't dare to do this because the market is imperfect now in China and it's difficult to supervise the private contractors through using contracts to ensure natural resources are not overused. See, [the] Siguniang Mountain case has demonstrated this way doesn't work'. If the JBRA had contracted out the tourism business in the reserve, it would have had to bear the high cost of supervising contracts and it would have carried heavy responsibility in terms of protecting natural resources from potential overuse. This is the second reason for the JBRA to select the joint-stock mechanism.

The discussion above does not explain why the JBRA decided to share the stocks with local residents instead of with external private companies like many other reserve administrators. With reference to the history of the JOC development, it was indicated that the joint-stock system was actually initiated by Shuzheng residents in 1990 to avoid the intense competition for tourists. It is posited, therefore, that the subsequent selection of a joint-stock company mechanism by the JBRA was path-dependent from the point of view of institutional evolution. During the questionnaire surveys, all the residents of Shuzheng felt satisfied with their innovative measures. They believed that the joint-operated family hotels brought in higher income by avoiding the intense competition for tourists, and it restored the traditional close relationships between households. From the JBRA viewpoint, joint operation was also considered as a positive mechanism because it improved the image of tourism in the JBR through a unified market price and service. Path-dependence theory (Mahoney 2000) proposes that people are inclined to continue with a course of action if they feel it is more positive than any alternative course. Therefore, path-dependence is the third reason for the JBRA to employ the joint-stock mechanism.

Another driving force behind the JBRA sharing stocks with local residents is that it is an organization implementing conservation on behalf of the government. While the organizational goal of the JBR is to conserve natural resources, its institutional orientation determines that it must consider benefits to the local community. The JBRA represents public interest in sustainable management of the natural resources within the reserve. If the local communities had not received benefits from tourism, then they would still be cultivating the steeply sloping fields and hunting in the reserve, resulting in possible ecological degradation. If the natural resources of the reserve were to be destroyed, the careers of the JBR administrators would be affected as they are appointed by the government.

In conclusion, the necessity of financial self-reliance (set against the backdrop of national economic policy); the potential problems in contracting out tourism business and stewarding the local environment; path-dependence; and the reserve management aim of linking conservation and development come together as the key reasons for applying the joint-stock system to tourism management in the JBR.

In order to develop viable economic businesses in nature reserves, theoretically the first action is to reform the resource management system, completely separating the reserve administration from business activities. Following the theory of new public management (Savas 2000), the government should only adopt the role of 'arranger', whilst the private sector or NGOs should act as the 'provider' by using contracts. It should be realized, however, that it will take much time to redefine the role of government, which might explain why many integrated conservation and development programmes (ICDPs) in developing countries have not been successful (Songorwa 1999, Twyman 2000, Brown 2002, Gupte 2003, Hutton and Leader-Williams 2003, Spiteri and Nepal 2006).

Given that the joint-stock operating mechanism is a relatively satisfactory way to balance local development and natural resource conservation under current institutional arrangements (albeit its suitability beyond the existing transitional period is unknown), this model appears to be a worthwhile reference for other protected areas, especially in those places where the conflict between local development and conservation is severe. The JBR case may provide an alternative way of thinking to solve such conflict, especially in developing countries.

Acknowledgements

Thanks are extended to the Natural Sciences Foundation of China for continuously supporting this research through three funded projects (49901005, 40271032 and 40671070). Thanks to all those students who helped to conduct surveys during the research. Finally, thanks must go to those indigenous people and managers of Jiuzhaigou Bioshere Reserve who gave us invaluable help during our survey.

References

Brown, K. (2002) Innovation for conservation and development. *The Geographical Journal*, 168, 617.

Burger, J. (2000) Landscape, tourism and conservation. *The Science of the Total Environment*, 249, 3949.

Collins, A. (1998) Tourism development and natural capital. *Annals of Tourism Research*, 26, 98109.

Gupte, M. (2003) Re-examining participatory environmental policy: Social stratification and the gender dimension. *Society and Natural Resources*, 16, 327334.

Hutton, J. and Leader-Williams, N. (2003) Sustainable use and incentive-driven conservation: Realigning human and conservation interests. *Oryx*, 37, 215226.

Kousis, M. (2000) Tourism and the environment: a social movements perspective. *Annals of Tourism Research*, 27, 468489.

Kuvan, Y. (2005) The use of forests for the purpose of tourism: The case of Belek Tourism Center in Turkey. *Journal of Environmental Management*, 75, 263274.

Laiolo, P. (2003) Diversity and structure of the bird community overwintering in the Himalayan subalping zone: Is conservation compatible with tourism? *Biological Conservation*, 115, 251262.

Li, W. (2006) Community participation in nature reserve ecotourism: Experiences from the Jiuzhaigou Biosphere Reserve, China. *Annals of Tourism Research*, 33, 132–143.

Li, W., Zhang, Q., Liu, C. and Xue, Q (2006) Tourism's Impacts on Natural Resources: a Positive Case from China. *Environmental Management*, 38, 572–579.

Lukashina, N.S., Amirkhanov, M.M., Anisimov, V.I. and Truner, A. (1996) Tourism and environmental degradation in Sochi, Russia. *Annals of Tourism Research*, 23, 654665.

Mahoney, J. (2000) Path dependence in historical sociology. *Theory and Society*, 29, 507–548.

Mbaiwa, J.E. (2003) The socio-economic and environmental impacts of tourism development on the Okavango Delta, north-western Botswana. *Journal of Arid Environments*, 54, 447467.

Nepal S.K. (2000) Tourism in protected areas: The Nepalese Himalayas. *Annals of Tourism Research*, 27, 661681.

Ross, S. and Wall, G. (1999) Ecotourism: Towards congruence between theory and practice. *Tourism Management*, 20, 123132.

Savas, E.S. (2000) *Privatization and Public-Private Partnerships*. Chatham House, New York.

Songorwa, A. (1999) Community-based wildlife (CWM) in Tanzania: are the communities interested? *World Development*, 27, 20612079.

SPAFS (Sichuan Provincial Academy of Forestry Sciences) (2004) *Comprehensive Scientific Survey Report of Jiuzhaigou Nature Reserve*. (In Chinese).

Spiteri, A. and Nepal, S.K. (2006) Incentive-based conservation programs in developing countries: a review of some key issues and suggestions for improvements. *Environmental Management*, 37, 1–14.

Twyman, C. (2000) Participatory conservation? Community-based natural resource management in Botswana. *The Geographical Journal*, 166, 323–335.

Van der Duim, R. and Caalders, J. (2002) Biodiversity and tourism: Impacts and interventions. *Annals of Tourism Research*, 29, 743761.

Zhang, X. (2005) The institutional flaws of world heritage and national key scenic sites managed in the way of separating rights bond. *Chinese Parks*, 7, 9–14. (In Chinese).

Zhang, X. (2006) Discussion on concession business applied in scenic sites and natural reserves. *Chinese Parks*, 1, 42–44. (In Chinese).

Zhang, X.Z. (2000) Natural and cultural heritage resource management and institutional reforms. *The Journal of Quantitative and Technical Economics*, 9, 9–13. (In Chinese).

Zhang, X.Z. (2002) Governance of national scenic sites. *Chinese Parks*, 2, 37–40. (In Chinese).

Chapter 6
Environmental Sustainability and Cruise Tourism in Arctic Canada

Emma J. Stewart and Dianne Draper

Introduction

The Arctic is facing unprecedented and momentous environmental, economic and socio-cultural change (Nuttall 1998, Nuttall and Callaghan 2000). Many commentators predict that as a consequence of a warming Arctic, maritime activity will increase, including the introduction of industrial fishing, expansion of non-renewable resource development and commercial shipping (Furgal and Prowse 2008). In combination, this increase in human activity in Arctic waters likely will add strain to an environment (both human and ecological) that already is stressed by the effects of climate change (Huebert and Yaeger 2007). There is a clear imperative to develop a broad framework in which new and existing maritime activities can be managed wisely, in perpetuity. Integrated management approaches are required so that development is equitable and environmentally responsible, protecting both the human and ecological environment of the Arctic (Berkes and Fast 2005, Huebert and Yaeger 2007).

Historically, ice conditions have precluded most commercial shipping in the Arctic, but in light of a warming climate, many Arctic specialists predict an increase in commercial shipping in this region (Huebert 2001, ACIA 2004, Brigham and Ellis 2004). As a form of commercial shipping, cruise tourism is predicted by some commentators to benefit from receding ice conditions, accelerating the modest growth already established in this niche travel sector (Huebert 2001, Scott 2003, ACIA 2004, Brigham and Ellis 2004, Huebert and Yaeger 2007). Although the high costs of ice-breaking once placed the Arctic off limits to cruise vessels (Ritter and Schafer 1998), the Arctic became a viable cruise tourism destination following the mid 1980s when the former Soviet Union's economy declined and its relatively inexpensive ice-breakers suddenly became available (Jones 1999, Stonehouse 2007). Since the MS *Explorer* first carried passengers to the Canadian Arctic in 1984, tourists have visited an increasing number of northern destinations (Marsh and Staple 1995, Jones 1999). Transits of the Northwest Passage are now commonplace; Baffin Island has been circumnavigated multiple times; and communities such as Resolute Bay, Pond Inlet and Pangnirtung are visited regularly. To date, fortunately, cruise operators in Arctic Canada have maintained

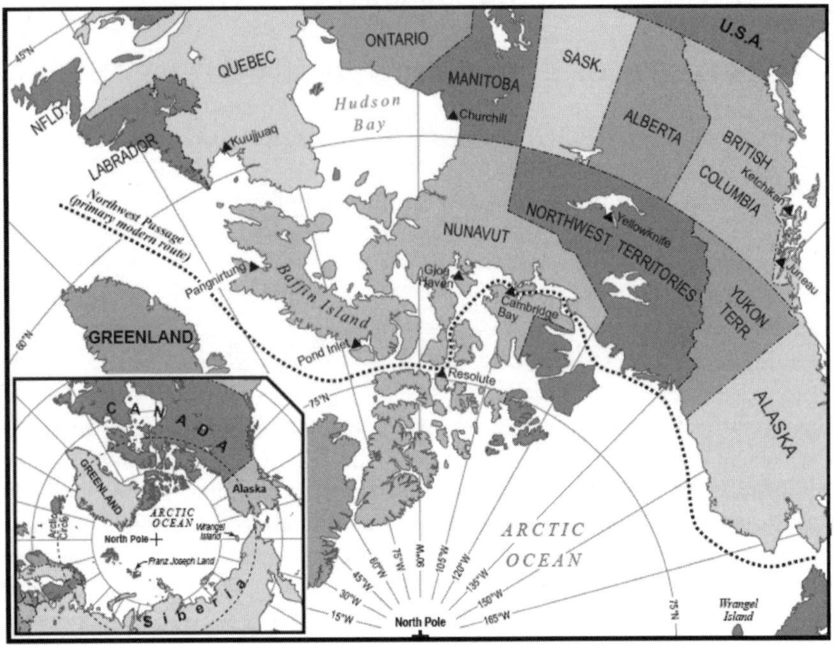

Figure 6.1 Selected cruise destinations and the route of the Northwest Passage in Arctic Canada

an excellent human safety record although there is a 'lengthy record and anecdotal history of groundings and other bumbles' (Jones 1999: 31).

This previously good environmental and human safety record was shattered in November 2007 when the polar veteran, MS *Explorer*, sank off the coast of the Antarctic Peninsula, an incident some commentators had been predicting for some time (Stewart and Draper 2006). The sinking of the MS *Explorer* brought into stark reality the individual, cultural and environmental risks associated with polar travel. Fortunately, human life was spared in this incident, but the environmental consequences of *Explorer*'s sinking remain unknown. It is clear that without adequate policy and integrated management and planning in place, what happened to the ice-strengthened MS *Explorer* in the Antarctic could happen to other ships operating in the Canadian Arctic (Stewart and Draper 2008). This is particularly so because, under climate warming, hull penetrating, multi-year ice may present navigational hazards for ships in some regions of the Canadian Arctic (Howell and Yackel 2004).

The purpose of this chapter is threefold. Firstly, the nature, scale and scope of cruise tourism to Arctic Canada are addressed, focusing particularly on what is known about environmental consequences of Arctic cruise tourism. Secondly, the way in which climate change may influence the development of cruise tourism in Arctic Canada is presented. Thirdly, we comment on the elements that likely will be required to ensure development of cruise tourism in northern Canada is both successful and sustainable for affected communities, the local environment, cruise operators and their clients (Stewart and Draper 2006). We do not offer new empirical evidence but rather, using desk-based literature searches, propose that existing approaches to managing cruise tourism activities should be replaced with broader, more integrated management approaches.

Whether or not the potential increase in cruise tourism in Arctic Canada is likely to be sustainable depends on how the term 'sustainable' is defined. In this chapter, the concept of sustainability developed by Berkes and Fast (2005) is appropriate. They suggest that sustainability is a process that includes ecological, soci-cultural and economic dimensions, whereby 'the needs of the present and local populations can be met without compromising the ability of future generations or populations in other locations [global in scope] to meet their needs' (Berkes and Fast 2005: 2). The question of 'what is to be sustained' is addressed on a case-by-case basis (Berkes and Fast 2005). We begin by establishing the current scale, scope and impact of the Canadian Arctic cruise industry and explore the evidence regarding its potential growth. We then focus on existing planning arrangements and present a rationale for an integrated approach that is mindful of the particular circumstances of cruise tourism in the north. An integrated oceans management approach is proposed as a means to identify more specific elements that communities, cruise operators and, to a lesser extent, their clients, can employ to fulfil sustainable management objectives. The emphasis is on collaboration and the need to build partnerships with key stakeholders of the marine environment. For the purposes of this chapter, Arctic Canada, as an area of study, is considered to be: northern Yukon, Northwest Territories, Nunavut, northern Manitoba, northern Quebec and northern Labrador (see Figure 6.1).

Scale and Scope of Cruise Tourism in the Canadian Arctic

The majority of cruise ships visiting the Canadian Arctic are retro-fitted icebreakers (such as the former Russian ship, *Kapitan Khlebnikov*) or ice-strengthened cruise vessels (such as the *Hanseatic*) (Figure 6.2). Expedition eco-cruising, a phrase used to describe the Arctic cruise concept, is styled on early cruises to Antarctica in the 1960s where cruising first combined brief shore visits by zodiacs (small inflatable craft) with environmental and historical education (Mason and Legg 1999, Splettstoesser 2000). The first expedition eco-cruise to the Canadian Arctic was made by the MS *Explorer* in 1984 (Marsh and Staple 1995, Jones 1999, Headland 2004). Carrying 98 passengers, the ship traversed the Northwest Passage from east to west in 23 days, only the thirty-third full passage ever; in doing so, the MS *Explorer* ushered in the Canadian Arctic cruise industry. Similar crossings were attempted, with only two successful traverses between 1984 and 1988 (Marsh and Staple 1995, Headland 2004). Despite these modest early developments, a more

Figure 6.2 *Hanseatic* cruise ship in Pond Inlet Nunavut

Source: Photographed by Emma J. Stewart.

regular pattern of cruise activity through the Canadian Arctic emerged from 1992 to 2005; for example, ships successfully traversed the Northwest Passage between one and three times per year.

In 2006, 22 cruise ships operated in the Canadian Arctic, double the number of ships in the previous season (Buhasz 2006). During the 2006 cruise season, communities on Baffin Island hosted as many as 12 cruise ships over a relatively short three month period; for example, 12 cruise ships docked at Pond Inlet over a 41-day period. Resolute, on Cornwallis Island, hosted 10 cruise ships. Resolute is an entry/exit point into or out of the Canadian High Arctic, and regularly scheduled air flights operate between Resolute, Cambridge Bay, Iqaluit and Ottawa. The community of Pangnirtung hosted four cruise ships; the hamlet of Arctic Bay hosted three cruise ships; and Grise Fiord, the most northerly community in Canada, located on the southern shore of Ellesmere Island, hosted three cruise ships (Stewart et al. 2007). Similar numbers of cruise vessels operated in the Canadian Arctic during the 2007 season. These trends suggest that the industry has moved beyond its infancy and is now entering a maturing phase with increased numbers of vessels, more regular and predictable patterns of activity, and new, more demanding routes (Stewart et al. 2007). A range of factors is likely to support this maturing phase of the industry, including increasing tourist demand for travel to remote places, overall popularity of cruising worldwide, more sophisticated promotional activities by tour agencies, and increasing awareness at the political and community levels about the potential benefits of cruise tourism (Dawson et al. 2007, Stewart et al. 2007, Dawson et al. in press).

Environmental Consequences of Cruise Tourism in the Canadian Arctic

There is very little research on the environmental effects of current levels of cruise activity in the Canadian Arctic, and only speculation about the possible effects presented by increased cruise traffic in the region. Despite a lack of specific research, but based on evidence presented elsewhere, cruise tourism is likely to have effects on various aspects of the Arctic maritime and terrestrial environments. In terms of the maritime environment, as with any form of shipping, there is always the threat of oil spills, which are difficult to clean up in cold remote locations (Huebert and Yaeger 2007, Marquez and Eagles 2007, World Wide Fund for Nature 2008). The possible introduction of invasive marine species through ballast water discharges is a hazard, as is the control of the ship's waste products (Huebert and Yaeger 2007). The risk of wildlife strikes and the impact of vessel activity in and around breeding grounds and feeding areas of marine mammals are also of concern (Gerrard 1999). Cruise vessels operating in remote locations, off the main shipping routes, must be able to respond appropriately in the face of inadequate emergency response capacity. In terms of the terrestrial environment, passengers make regular shore visits to communities and other places of natural or historical interest as part of their cruise, either by zodiac or by helicopter. Some concern has been raised about the consequence of cruise tourists visiting archaeological sites (because of the potential for unintended trampling) as well as noise disturbance caused by helicopter activity (Marquez and Eagles 2007). There is potential for disturbance to Arctic flora and fauna since these species have a low tolerance to impacts that ecosystems in more temperate latitudes would absorb more easily (Snyder 2007). There is similarly the potential for social and cultural disruption among indigenous peoples whose communities play host to cruise visitors on an increasingly regular basis (Nuttall 2005).

While Arctic cruise tourism has the potential to disrupt local communities, it also has the potential to enrich them. Some proponents suggest that tourism can contribute positively to the economies of remote communities (Jones 1999), while others have suggested that tourism to the Polar Regions, in general, has the potential to create ecologically-minded ambassadors (Maher et al. 2003a, 2003b). Such positive perspectives lead some commentators to suggest that cruise tourism in the Arctic can be viewed as an environmentally sustainable option (Ritter and Schafer 1998). Still others dispute this, claiming that problems with waste generation and disposal, together with emissions from long-haul flights and cumulative pressures on fragile environments and host communities, are likely to

Table 6.1 Some aspects of global environmental change with relevance for Arctic tourism

Effects on infrastructure
- Increased open water leads to increased storm surges and shoreline erosion
- Permafrost melting/land instability leads to construction and engineering problems and structural damage

Effects on access
- Decline in sea-ice extent leads to extended shipping season
- Glacier melting leads to increased iceberg hazards
- Shorter seasonal river ice duration leads to access difficulties related to winter roads
- Earlier and greater spring floods leads to access hazards
- Greater snow accumulation leads to access difficulties
- Northward movement of permafrost line leads to increased access through road construction

Effects on attractions
- Greater snow accumulation leads to new opportunities for snow-based activities
- Shorter snow duration leads to seasonal challenges for some activities
- Warmer summer and winter temperatures lead to extension of seasonal activities
- Warmer summer temperatures leads to increased insect challenge
- Warmer winter temperatures lead to new opportunities
- Ecosystem changes lead to alterations in distribution and abundance of existing animal species
- Ecosystem changes lead to appearance of new species in north
- Environmental changes alter local activity possibilities
- Scenic values altered through environmental changes locally and regionally

Source: Table 2.1, Johnston M.E. (2006) Impacts of global environmental change on tourism in the Polar Regions. In S. Gossling and C.M. Hall (eds) *Tourism and Global Environmental Change: Ecological, Social, Economic and Political Interrelationships* Routledge, New York, p. 44.

make cruise tourism far from sustainable in this region (Marsh and Staple 1995). This debate is complicated by the environmental, socio-cultural and economic consequences of a warming Arctic.

Climate Change and Arctic Cruise Tourism

It is now well established that the Canadian Arctic is a region particularly sensitive and vulnerable to changes in the global climate system (ACIA 2004, IPCC 2007). There is overwhelming evidence to suggest that glaciers are melting, sea ice is receding (evidenced in the summer of 2007 when the Northwest Passage was ice free), permafrost is melting, ranges of flora and fauna are shifting and fires and insect outbreaks are increasing (ACIA 2004). Specialists in climate change in the Arctic region suggest that:

- Arctic climate is now warming rapidly and much larger changes are projected;
- Arctic warming and its consequences have worldwide implications;
- Arctic vegetation zones are very likely to shift causing wide-ranging impacts;
- Animal species' diversity, ranges and distribution will change;
- Many coast communities and facilities face increasing exposure to storms;
- Reduced sea ice is very likely to increase marine transport and access to resources;
- Thawing ground will disrupt transportation, buildings and other infrastructure;
- Indigenous communities are facing major economic and cultural impacts;
- Elevated ultraviolet radiation levels will affect people, plants and animals;
- Multiple influences interact to cause impacts to people and ecosystems.

(ACIA 2004: 10–11)

The extent to which cruise tourism will be affected by such environmental change in Arctic Canada is mostly unknown and largely a matter of conjecture. Pagnan (2003) speculates Canada might expect larger cruise ships such as those that operate along the Alaskan coast, and now in Antarctica (see: Bertram 2007) while others also point to potential increases in the number of small adventure yachts. According to the Joint Task Force North (JTFN) six yachts were in Canadian waters in 2006 and this trend is predicted to increase, as it has in Antarctica in recent years (see: Bertram 2007). Johnston (2006) suggests that climate change will present both opportunities and challenges for Arctic tourism in terms of infrastructure, access and activities (see Table 6.1).

The retreat of Arctic sea ice and increased ice hazards to shipping are perhaps the most relevant aspects of environmental change with regard to cruise tourism. Stewart et al. (2007) explored the specific relationship between changing sea ice conditions and cruise travel in the Canadian Arctic, concluding that global climate model predictions of an ice free Arctic as early as 2025 may present a false sense of optimism about the potential for a blanket increase in cruise traffic throughout the Canadian Arctic. This is because under climate change, regions such as the High Arctic and Northwest Passage may experience heavier than normal volumes of multi-year ice, which could present navigational risks to cruise vessels. By contrast, in other areas such as Baffin Bay, increasing open water could promote safer travelling conditions which are more attractive to cruise operators. However, with the absence of ice comes the reduction in opportunities for tourists to view ice-dependent wildlife such as polar bears, seals and walruses. Evidence from cruise passengers visiting Arctic Canada suggests that for many tourists the opportunity to see such wildlife species 'before they were gone forever' was a key factor in their decision to visit (Maher and Meade 2008). Tour operators may be able to capitalise on 'last chance to see' tourism in the short term, but clearly this is environmentally unsustainable in the long term.

Regardless of what happens to sea ice and the animals dependent on it for survival, the Canadian Arctic likely will see moderate increases in the volume of cruise traffic, at least in the short term. Latent demand and the propensity for an increasing number of cruise visitors to become return patrons form the basis of this prediction. Even in the early days of northern cruising 30 per cent of tourists indicated they would return to the Arctic (Marsh and Staple 1995), but whether those tourists returned to the north is unknown. Research in Antarctica confirms that people who have visited 'one polar region are also likely to want to experience the other' (Bauer 2001: 153). Such trends suggest the potential for modest growth in Arctic cruise tourism; modest rather than rapid growth is predicted because growth is limited by the current availability of ships equipped for polar travel. An important observation is that large, non-ice-strengthened passenger liners have been in operation in Antarctica since 2000 (Bertram 2007), heralding a trend that may extend to the polar north if ice conditions are favourable.

Existing Planning Mechanisms

Despite recent and predicted growth there has been little coordinated, trans-regional planning for the sustained development of cruise tourism in Arctic Canada. This is not surprising given the complex interaction of multi-scale stakeholders involved in cruise tourism in this region. As some commentators have suggested the labyrinth of regulation generated from the different levels of governance (from local through to international) is complex and unnecessary (Marquez and Eagles 2007). For example, at a local level, Parks Canada, Environment Canada, Canadian Wildlife Service and Hamlets are involved in providing tourists with permits for shore visits; at a regional level Nunavut Tourism, the Government of Nunavut, Customs Canada, Transport Canada, Department of Fisheries and Oceans and the Coast Guard are all active in overseeing the development and safety of the industry as well as being responsible for implementation and monitoring of various levels of existing shipping regulations; and at an international level the Arctic Council and the Association of Arctic Expedition Cruise Operators (AECO) work to ensure that voluntary codes of practice are adhered to, although Arctic Canada falls outside of the regional remit of AECO (see: AECO 2008).

Concerns over sovereignty in Canadian Arctic waters further complicate an already complex management scenario (see: Huebert 2001, Huebert and Yaeger 2007), leaving the impression that cruise tourism is either a non-issue or is left to the behest of the cruise industry. Neither situation is helpful or acceptable; the real challenge is to determine how to do things differently. This situation concurs with observations by Hall (2001: 614) that 'planning agencies often fail to understand tourism, while tourism promotion authorities tend not to be involved with the evaluation of its effects or its planning and management. Implementation strategies often fail to recognize the interconnectedness that exists between agencies in trying to manage environmental issues particularly when, as in the case of the relationship between tourism and the environment, responsibilities may cut across more traditional lines of authority'.

An Integrated Coastal Management Approach

While stakeholder involvement in tourism development generally is common practice (Lester and Weeden 2004), the importance and role of stakeholders in the planning and development of cruise tourism in the north is just beginning to be recognized. For example, Johnson (2002: 268) concluded that the industry needed to 'continue to take a long-term view, fostering holistic integrated management planning involving international agencies, cruise line operators and host communities'. Mindful of the inherent difficulties of collaborative practice, we propose that cruise operators, communities and other relevant stakeholders might benefit from cruise tourism planning being given higher priority in the broader framework of integrated coastal management.

Integrated management in the coastal environment was ushered in during the 'late 1980s when it became clear that the ocean and coastal resources were not sustainable under the conventional approaches of managing single activities and species' (Berkes and Fast 2005: 8). Integrated management is defined as 'a continuous process through which decisions are made for the sustainable use, development and protection of areas and resources' (Canada's Oceans Strategy 2002: 36). An integrated approach may be particularly useful in the north where resource management operates in an environment of uncertainty (Berkes and Fast 2005). The utility of integrated management in the north occurs because this approach: acknowledges the existing interrelationships among different uses, users and the environments they potentially affect; overcomes the fragmentation inherent in existing sectoral management approaches; analyzes the implications of development and conflicting uses; and promotes linkages and harmonization among various activities (Canada's Oceans Strategy 2002). Against a backdrop of environmental change, making decisions in an integrated fashion might enable managers to address concerns about: the variable nature of cruise tourism across the Canadian north; the volume and frequency of visitors to small Arctic communities; and the wide-ranging effects of tourism on the marine and terrestrial Arctic environments.

Integrated management is one of the three principles on which Canada's Oceans Strategy (2002), the implementation mechanism of the Oceans Act (1997), was constructed (Oceans Act 1997, Berkes and Fast 2005). Given that Canada was the first country in the world with comprehensive oceans management legislation, it comes as some surprise that tourism is not recognized in either the Act or the Strategy. Although tourism currently is unacknowledged in much of the rhetoric of Canada's oceans management policy, it would seem reasonable to suggest that key stakeholders involved in northern cruise tourism might align themselves to the existing integrated coastal management framework.

Table 6.2 Lisbon Principles for sustainable ocean governance

- *responsibility principle* stating that it is the responsibility of individuals or corporations to use environmental resources in an ecologically sustainable, economically efficient and socially just manner
- *subsidiarity principle* emphasizing the importance of local-level institutions in governance
- *adaptive management principle* highlighting the requirement to continually monitor and adapt systems
- *precautionary principle* stating that in cases where there are threats of serious or irreversible damage, lack of full scientific certainty shall not be used as a reason for postponing cost-effective measures to prevent environmental degradation
- *participation principle* pointing to the importance of full stakeholder participation in the formulation and implementation of decisions about environmental resources

Source: Berkes and Fast (2005)

Guiding Principles

Not only would an integrated coastal management framework situate tourism in a wider context that views marine activities holistically, it would provide an important set of guiding principles, including the responsibility, subsidiarity, adaptive management, precautionary and participation principles, known as the Lisbon Principles for sustainable ocean governance (Berkes and Fast 2005) (see Table 6.2).

However, the framework does not provide specific elements that are required to translate well-intended principles into meaningful practice. In order to address these gaps, Table 6.3 illustrates a range of possible strategies and management initiatives that communities, operators, and, to a lesser extent, tourists, could employ to fulfil sustainability obligations. While many cruise tourism operators might believe they already embrace these somewhat idealistic principles, it is useful to restate them here because they are helpful in guiding the future development of cruise tourism in Arctic Canada. For instance, these principles emphasize the responsibility on tour operators to use environmental resources wisely; that given uncertainty surrounding future levels of tourism impact, this should not delay the implementation of measures to ensure environmental protection; that there is a need to monitor the cruise tourism industry and re-evaluate and revise management actions on a continual basis; that decision-making should be made by the most capable local organisations; and that collaboration between aboriginal land claim and other agencies, coastal communities, tour operators, conservation agencies and other stakeholders in the marine environment, should underpin the whole process.

Table 6.3 Cruise tourism within an Integrated Oceans Management Framework

Integrated Oceans Management Framework				
Overarching Principles				
Responsible	**Subsidiary**	**Adaptive**	**Precautionary**	**Participatory**

Cruise Tourism in Arctic Canada: Possible Integrated Management Strategies and Initiatives			
Strategy	**Community**	**Operator**	**Tourist**
Environmental	• Limit visitor numbers • Create a 'shore plan' • Determine limits to acceptable change	• Vary itineraries • Limit passenger numbers • Tackle cumulative environmental impacts • Implement environmental policy	• Adhere to codes of conduct • Report inappropriate activity • Practice appropriate photography
Economic	• Promote local crafts and products • Consider 'head taxes' to off-set costs	• Environmental cost accounting • Invest in clean technology • Reward environmental awareness	• Contribute to local enterprise as far as possible during all shore visits
Socio-Cultural	• Establish limits to community visits (number and timing), *i.e.* carrying capacity • Promote local culture	• Promote appropriate community visit behaviour • Implement code of conduct	• Be mindful of behaviour and sensitive to local people and their lifestyles
Educational	• Provide appropriate combination of interpretive and educational material	• Liaise with appropriate community contacts at all stages of planned shore visits	• Be well informed prior to shore visits
Political	• Gain support for development from government agencies • Develop framework for management and development	• Work with key stakeholders to ensure that best practice informs policy	• Engage i[illegible] upon ret[illegible]
Research			

Source: After Johnson (2002)

ecific Elements

Table 6.3 presents a number of specific integrated management elements in the form of strategies and management initiatives that communities, tour operators and tourists could (and perhaps should) implement. Communities could adopt a series of initiatives to help mitigate possible negative effects of cruise visitation. For instance, communities may wish to determine the upper limits of passenger numbers who disembark, thereby establishing a destination carrying capacity. Some commentators suggest that in many cruise destinations carrying capacity has been ignored, although a few attempts to set capacity limits have been made (Johnson 2002, Lester and Weeden 2004). For example, in Antarctica the New Zealand Heritage Trust has set carrying capacities for each of the historic huts under its management in the Ross Dependency (Hughes and Davis 1995). Clearly a community benefits by identifying how much change its residents are prepared to accept given likely increases in cruise tourism. This necessarily involves identifying how much tourism is too much, both from an environmental and a cultural stance.

Concern about over-capacity has prompted some cruise destinations to introduce 'head taxes' (Johnson 2002). For example, Bermuda imposes the highest head tax of $US63 per passenger for all tourism vessels, the Bahamas charges $US15 and residents in Juneau, Alaska are debating a $US5 head tax to off-set costs associated with environmental impacts of cruise tourism vessels. Unfortunately, the legitimacy of this practice may be challenged as it is difficult to verify the application of these taxes to environmental amelioration (Johnson 2002). Other strategies and initiatives for communities could include promotion and careful coordination of sales of local products and crafts; consideration of how communities might appropriately involve local culture and indigenous knowledge in shore visits; development of particular interpretive approaches to ensure that pre-determined messages are communicated effectively to ship passengers; and consideration of how community members could get involved in developing policy.

Similarly, operators could (and perhaps should) consider the possibility of varying itineraries to ensure that there is some equity in visitation across the many varied communities of the Canadian Arctic. Certain communities already appear to be emerging as favoured destinations in the Canadian north (for example, Pond Inlet, Pangnirtung and Cape Dorset). This is also the case in Antarctica where honey-pot sites have emerged (Mason and Legg 1999). Although evidence from elsewhere suggests it is unlikely that operators determine cruise itineraries on the basis of sustainability issues, it is important for operators to consider hitherto less popular destinations and include them in addition to, rather than instead of, well established Canadian Arctic destinations (Johnson 2002). Achieving this goal will demand careful planning and negotiation at all levels to enable a greater distribution of economic benefits among Arctic communities, whilst distributing tourists more widely across the environment. Experience elsewhere, however, cautions that 'impervious behaviour by cruise lines … seriously compromise[s] the actuality of equity amongst all stakeholders in managing the growth and development of cruise tourism' (Lester and Weeden 2004: 46).

Tour operators could appoint a designated person to liaise with appropriate community contacts (and Nunavut Tourism, Tourism Industry Association of the Yukon, Northwest Territories Tourism) and remain in contact at all stages of the planned shore visit, from the early stages of planning through to arrival, shore visit and departure (Marquez and Eagles 2007). Also, tour operators immediately could implement voluntary codes of conduct thereby promoting acceptable behaviour among their passengers whilst in communities. This has been addressed partially in the European Arctic, where the Association of Arctic Expedition Cruise Operators (AECO) has established voluntary guidelines for expedition cruising (see: AECO 2008). Some of the members of AECO already operate in the Canadian Arctic, so presumably these guidelines are adhered to by at least some cruise operators working in this region. Similar codes of conduct have been developed by the International Association of Antarctica Tour Operators (IAATO) where visitors on board IAATO member expeditions are reminded, for example: not to disturb wildlife, not to walk on fragile plants, and not to interfere with protected areas or scientific research (Bauer and Dowling 2003, IAATO 2005). However, these codes are voluntary and are self-implemented by tour operators with little or no way of policing visitor, or tour operator, behaviour. With the exception of research by Davis (1995, 1999) and Grenier (1998, 2003, 2004) little is known about actual visitor behaviour once on shore and whether or not such guidelines are adhered to strictly.

Tour operators also could work with a variety of coastal stakeholders to develop policy that is meaningful, practical and sensitive to the northern context. IATTO may be a good model as their aims include advocating, promoting and practising environmentally responsible private-sector travel to Antarctica (Splettstoesser 2000). While tour operator membership in IAATO is voluntary, since 1991, members claim to have established procedures and codes of conduct that ensure appropriate, safe and environmentally sound travel to the Antarctic (IAATO 2005). IAATO cooperates closely with governments and researchers and has attended relevant Antarctic Treaty Consultative Meetings (ATCM) illustrating that tour operators can engage effectively in the wider policy process. Although not a straightforward exercise, Arctic-bound cruise operators could investigate the necessary conditions for extending IAATO to cover northern Polar Regions, including Arctic Canada (Geitz 2005, Marquez and Eagles 2007).

As tourists also have a duty of care for the environment, passengers should: adhere to codes of conduct despite the fact that they are voluntary; engage in non-intrusive and appropriate photography (both of the natural and cultural environments); contribute (even if a small amount) to each community visited to ensure equity of spending (where possible); take it upon themselves to be well informed about the places they visit before they disembark; and, finally, engage in ambassadorial activities. In Antarctica, there are claims, difficult to prove and so far unsubstantiated, that Antarctic tourists are the best possible ambassadors for the

Antarctic, as they spread the message about the need for environmental protection when they return home (Maher et al. 2003a). However, others caution that tourists also become important marketing tools where 'the accounts of their travels might persuade new potential visitors to follow in their footsteps' (Grenier 2004: 424), even if those visitors have not 'trodden lightly'. For some commentators, therefore, the use of the label environmental ambassadors for these visitors is problematic.

A final key element, which underscores the ICM framework, is the need to ensure that decision-makers are cognisant of the research that has been undertaken on tourism, its management, planning and effects, as well as identifying areas where there are research gaps. High impact scholarship needs to be pursued, that is, research that satisfies our deep curiosity, solves practical problems and contributes to public policy (Harman 2003). In the Arctic, this means that local and regional interests must be a foundational component of tourism research (Stewart et al. 2005). Local communities as well as national ones will need to have a say in how priorities are developed. Indeed, partnership in research is an absolute necessity in the North (Watt-Cloutier 2000).

Although we have focused largely on Arctic Canada here, it is critical to adopt an Arctic-wide approach, so that there is consistency and cooperation across the eight Arctic nations. Policies should cross-cut regional and national boundaries to reflect the tour itineraries of cruise ships. While acknowledging inherent difficulties, this approach is especially important given the uncertainty about Russia's role in the future of cruise tourism. Russia occupies about half of the world's Arctic territory and owns the majority of ice-breaking vessels, so their involvement in the development of the industry is paramount (Jones 1999). Some commentators have called for a regional treaty system, akin to the Antarctic Treaty System (ATS), to provide a suitably robust framework for more holistic Pan-Arctic management, as well as raising the profile of the Arctic on domestic as well as on international stages (Huebert and Yaeger 2007). A strengthened Arctic Council, as well as more binding marine regulations, could be important steps to securing comprehensive management of the Arctic in the future, which would include new economic activities such as cruising (Huebert and Yaeger 2007).

The elements presented in Table 6.3 already may be in existence in some situations, but not in others. While recognizing the realities of working in the field, in the ideal world, we intend the ideas listed in Table 6.3 as a series of options that need to be assessed on a case-by-case basis and not as exhaustive or exclusive items. However, experience suggests that collaborative efforts have been 'short lived and ultimately unsuccessful' (Lester and Weeden 2004: 45) and often have been compounded by specific challenges such as the heterogeneous nature of destinations, different assessments of the nature of problems facing cruise destinations, and the many different stakeholder interests (Lester and Weeden 2004). So, it is only with considerable effort that these challenges will be addressed in the Canadian Arctic context. By adhering to the guiding principles, by cherry-picking appropriate strategies and by engaging in and being aware of appropriate research, real progress may be made in ensuring that northern cruise tourism is both successful and sustainable environmentally, socio-culturally and politico-economically. By maintaining the current status quo it is likely that tour operators will continue to lead development and, in the absence of an Arctic-wide equivalent of IAATO, this might result in an uneven and ultimately unsustainable industry. If managed appropriately and responsibly, northern cruise tourism can facilitate the fostering of meaningful encounters between Arctic hosts and guests, can make possible the engagement in rich learning and can boost the economies of northern Canadian communities whilst also ensuring the maintenance of ecological processes.

Conclusion

While the cruise industry is a potential vehicle for the development of ecotourism in Arctic Canada, evidence from elsewhere suggests that there is a need to take a long term view toward adoption of holistic integrated planning approaches; for operators to continue to invest in good environmental practice; for political will to safeguard destinations; for greater profit sharing among shareholders and destination communities; and for both operators and communities to raise environmental awareness and practice of environmentally responsible activities (Johnson 2002). To ensure that these lessons are applied to the cruise industry in northern Canada, we proposed a series of interrelated elements, including an oceans integrated framework coupled with guiding principles, and a variety of strategies and initiatives designed to meet the needs of particular situations on a case-by-case basis. The intention is to raise the profile of cruise tourism, moving it from its apparent current status of being a non-issue to one which recognizes its inevitability and growing importance in the Canadian Arctic.

The current management of cruise tourism in Arctic Canada is not sufficiently robust to deal with the current impacts of cruise tourism, let alone plan for anticipated changes as a result of climate warming. There is a sense of urgency to address the issues raised here because changes ushered in by new economic activities and climate change likely will accelerate the development of cruise tourism in some regions of the Canadian Arctic. While cruise operators in Arctic Canada possess an excellent human safety record, as the sinking of the MS *Explorer* in 2007 illustrated the Polar Regions are places of danger, and accidents can completely change the development of the industry in an instant. It is crucial that decision-makers take proactive steps to ensure that individual, cultural and environmental risk is minimized. This is an important platform upon which local people, tour operators, and other interested stakeholders can develop an industry which is a viable and appropriate vehicle for sustainable tourism in the north, and one that provides visitors with the privilege of experiencing the beautiful and challenging environment of Arctic Canada.

Acknowledgements

An earlier version of this chapter appeared in the journal *Tourism in Marine Environments* and the authors are grateful to Micha Lück for his advice on the development of this work. Emma wishes to thank The Pierre Elliot Trudeau Foundation for her doctoral scholarship. And thanks go to Robin Poitras, Cartographer in the Department of Geography at the University of Calgary, for creating the map used in this chapter.

References

ACIA (2004) *Arctic Climate Impact Assessment: Impacts of a Warming Arctic*. Cambridge University Press, Cambridge.

AECO (2008) *Association of Arctic Expedition Cruise Operators*. Available at: http://www.aeco.no/ [accessed: 8 March 2008].

Bauer, T. (2001) *Tourism in the Antarctic: Opportunities, Constraints and Future Prospects*. The Haworth Hospitality Press, London.

Bauer, T. and Dowling, R. (2003) Ecotourism Policies and Issues in Antarctica. In D. Fennell and R. Dowling (eds) *Ecotourism Policy and Planning*. CABI Publishing, Wallingford, pp. 309–329.

Berkes, F. and Fast, H. (2005) Introduction. In F. Berkes, R. Huebert, H. Fast, M. Manseau and A. Diduck (eds) *Breaking Ice: Renewable Resource and Ocean Management in the Canadian North*. University of Calgary Press, Calgary, pp. 1–19.

Bertram, E. (2007) Antarctic ship-borne tourism: an expanding industry. In J.M. Snyder and B. Stonehouse (eds) *Prospects for Polar Tourism*. CABI Publishing, Wallingford, pp. 149–169.

Brigham, L. and Ellis, B. (eds) (2004) *Proceedings of the Arctic Marine Transport Workshop*, 28–30 September 2004, Scott Polar Research Institute, Cambridge University, Cambridge.

Buhasz, L. (2006) Northern underexposure. *Globe & Mail*, 1 July.

Canada's Oceans Strategy (2002) *Our Oceans, Our Future: Policy and Operational Framework for Integrated Management of Estuarine, Coastal and Marine Environments of Canada*. Fisheries and Oceans Canada, Ottawa.

Davis, B. (1995) Antarctic visitor behaviour: are guidelines enough? *Polar Record*, 31, 327–334.

Davis, B. (1999) Beyond guidelines: a model for Antarctic tourism. *Annals of Tourism Research*, 26, 516–533.

Dawson, J., Maher, P.T. and Slocombe, D.S. (2007) Climate change, marine tourism and sustainability in the Canadian Arctic: contributions from systems and complexity approaches. *Tourism in Marine Environments*, 4, 69–83.

Dawson, J., Stewart, E.J., Slocombe, S. and Maher, P. (in press) Climate change, complexity and cruising in Canada's Arctic: a Nunavut case study. In R.B.

Anderson and R.M. Bone (eds) *Natural Resources and Aboriginal People in Canada*. Captus Press, Concord.

Furgal, C. and Prowse, T.D. (2008) Northern Canada. In D.S. Lemmen, F.J. Warren, J. Lacroix and E. Bush (eds) *From Impacts to Adaptation: Canada in a Changing Climate*. Government of Canada, Ottawa, pp. 57–118.

Geitz, M. (2005) *Cruise tourism in the Arctic*. Available at: http://www.lighthouse–foundation.org/lighthouse–foundation.org/eng/forum/artikel00433eng.html [accessed 28 October 2005].

Gerrard, C. (1999) *Dugong-Watching Tourism and Encounter Response of the Dugong (Dugong dugon) in Shark Bay, Western Australia*. Unpublished MSc thesis, University of Calgary.

Grenier, A.A. (1998) *Ship-Based Polar Tourism in the Northeast Passage: A Case Study*. University of Lapland, Rovaniemi.

Grenier, A.A. (2003) *Croisières et tourisme polaire dans le Passage du Nord-Est*. University of Lapland, Rovaniemi.

Grenier, A.A. (2004) *The Nature of Nature Tourism*. University of Lapland, Rovaniemi.

Hall, C.M. (2001) Trends in ocean and coastal tourism: the end of the last frontier? *Ocean and Coastal Management*, 44, 601–618.

Harman, J.R. (2003) Whither Geography? *The Professional Geographer*, 55, 415–421.

Headland, R. (2004) Northwest Passage voyages. In L. Brigham and B. Ellis (eds) *Arctic Marine Transport Workshop*. Scott Polar Research Institute, Cambridge University, pp. A19–25.

Howell, S.E.L. and Yackel, J J. (2004) A vessel transit assessment of sea ice variability in the western Arctic, 1969–2002: implications for ship navigation. *Canadian Journal of Remote Sensing*, 30, 205–215.

Huebert, R. (2001) Climate change and Canadian sovereignty in the Northwest Passage. *Isuma: Canadian Journal of Policy Research*, 2, 86–94.

Huebert, R. and Yaeger, B. (2007) *A New Sea: The Need for a Regional Agreement on Management and Conservation of the Arctic Marine Environment*. WWF International Arctic Programme, Oslo.

Hughes, J. and Davis, B. (1995) The management of tourism at historic sites and monuments. In C.M. Hall and M.E. Johnston (eds) *Polar Tourism: Tourism in the Arctic and Antarctic Regions*. John Wiley & Sons, Chichester, pp. 235–255.

IAATO (2005) International Association of Antarctica Tour Operators. Available at: www.iaato.org [accessed 28 October 2005].

IPCC (2007) *Climate Change 2007: The Physical Science Basis: Summary for Decision Makers*. IPCC, Paris.

Johnson, D. (2002) Environmentally sustainable cruise tourism: a reality check. *Marine Policy*, 26, 261–270.Johnston, M.E. (2006) Impacts of global environmental change on tourism in the Polar Regions. In S. Gossling and C.M. Hall (eds) *Tourism and Global Environmental Change: Ecological,*

Social, Economic and Political Interrelationships. Routledge, New York, pp. 37–53.

Jones, C.S. (1999) Arctic ship tourism: an industry in adolescence. *The Northern Raven*, 13, 28–31.

Lester, J. and Weeden, C. (2004) Stakeholders, the natural environment and the future of Caribbean cruise tourism. *International Journal of Tourism Research*, 6, 39–50.

Maher, P.T. and Meade, D. (2008) *Cruise Tourism in Auyuittuq, Sirmilik, and Quttinirpaaq National Parks: Project Technical Report* (2007 season) – Draft. Parks Canada, Iqaluit.

Maher, P.T., Steel, G. and McIntosh, A. (2003a) Antarctica: tourism, wilderness and 'ambassadorship'? In A. Watson and J. Sproull (eds) *Science and Stewardship to Protect and Sustain Wilderness Values*. USDA Forest Service, Rocky Mountain Research Station, Port Elizabeth, South Africa, pp. 204–210.

Maher, P.T., Steel, G. and McIntosh, A. (2003b) Examining the experiences of tourists in Antarctica. *International Journal of Tourism Research*, 5, 59–67.

Marquez, J.R. and Eagles, P.F.J. (2007) Working towards policy creation for cruise ship tourism in parks and protected areas of Nunavut. *Tourism in Marine Environments*, 4, 85–96.

Marsh, J. and Staple, S. (1995) Cruise tourism in the Canadian Arctic and its implications. In C.M. Hall and M.E. Johnston (eds) *Polar Tourism: Tourism and the Arctic and Antarctic regions*. John Wiley & Sons, Chichester, pp. 63–72.

Mason, P.A. and Legg, S.J. (1999) Antarctic tourism: activities, impacts, management issues and a proposed research agenda. *Pacific Tourism Review*, 3, 71–84.

Nuttall, M. (1998) *Protecting the Arctic: Indigenous Peoples and Cultural Survival*. Harwood Academic, Amsterdam.

Nuttall, M. (2005) Tourism. In M. Nuttall (ed.) *Encyclopaedia of the Arctic* (Vol. 3 (O–Z)). Routledge, New York, pp. 2037–2040.

Nuttall, M. and Callaghan, T.V. (eds) (2000) *The Arctic: Environment, People and Policy*. Harwood Academic, Amsterdam.

Oceans Act, S.C.1996, C.31. In force 31 January 1997.

Pagnan, J.L. (2003) *Climate Change Impacts on Arctic Tourism*. Paper presented at the 1st International Conference on Climate Change and Tourism, 9–11 April, Djerba, Tunisia.

Ritter, W. and Schafer, C. (1998) Cruise-tourism: a chance of sustainability. *Tourism Recreation Research*, 23, 65–71.

Scott, D. (2003) *Climate Change and Tourism in the Mountain Regions of North America*. Paper presented at the 1st International Conference on Climate Change and Tourism, 9–11th April, Djerba, Tunisia.

Snyder, J.M. (2007) *Tourism in the Polar Regions: The Sustainability Challenge*. United Nations Environment Programme, Paris.

Splettstoesser, J. (2000) IAATO's stewardship of the Antarctic environment: a history of tour operator's concern for a vulnerable part of the world. *International Journal of Tourism Research*, 2, 47–55.

Stewart, E.J. and Draper, D. (2006) Sustainable cruise tourism in Arctic Canada: an integrated coastal management approach. *Tourism in Marine Environments*, 3, 77–88.

Stewart, E.J. and Draper, D. (2008) The sinking of the MS *Explorer*: implications for cruise tourism in Arctic Canada. *Arctic (InfoNorth)*, 61, 224–231

Stewart, E.J., Draper, D. and Johnston, M.E. (2005) A review of tourism research in the Polar regions. *Arctic*, 58, 383–394.

Stewart, E.J., Howell, S.E.L., Draper, D., Yackel, J. and Tivy, A. (2007) Sea ice in Canada's Arctic: implications for cruise tourism. *Arctic*, 60, 370–380.

Stonehouse, B. (2007) Tourism and the polar environment: Introduction. In J.M. Snyder and B. Stonehouse (eds) *Prospects for Polar Tourism*. CABI Publishing, Wallingford, pp. 1–14.

Watt-Cloutier, S. (Writer) (2000) Speech given on August 23. Silarjualiriniq 5, 2–5.

World Wide Fund for Nature (2008) WWF calls for moratorium on oil exploration in the Arctic. Available at: http://www.wwf.ca/NewsAndFacts/NewsRoom/default.asp?section=archive&page=display&ID=1573&lang=EN [accessed 17 March 2008].

Chapter 7
Managing Bear-Viewing to Minimize Human Impacts on the Species in Alaska

Terry D. DeBruyn and Tom S. Smith

Introduction

The United States Fish and Wildlife Service has conducted the National Survey of Fishing, Hunting, and Wildlife-Associated Recreation at five-year intervals since 1955. This survey has been recognized as the most instructive data set regarding fish- and wildlife-related recreation in the United States (US) (Grambsch and Fisher 1989). The 2006 survey revealed that the number of people hunting and fishing across the US has significantly declined. Since 2001, the number of people fishing and hunting in the US dropped by 12 per cent and 4 per cent respectively. During this same time period, declines in Alaska were even more pronounced, with decreases of 26 per cent in fishing and 24 per cent in hunting. But while the number of people hunting and fishing decreased, the survey showed that the number of people engaged in wildlife viewing increased. Nationwide, wildlife viewers increased 8 per cent from 2001 to 2006, whereas in Alaska, it jumped an impressive 22 per cent. Furthermore, the survey showed that, in 2006, 514,000 adults participated in Alaskan wildlife viewing and spent over $705 million (US Fish and Wildlife Service 2007).

Opportunities for bear-viewing in Alaska are numerous. There are approximately 30,000 to 40,000 brown bears in Alaska (Miller and Schoen 1999), making up 70 per cent of the brown bears in North America, or 98 per cent of the brown bears in the US. While no official state-wide estimate is available, experienced Alaskan wildlife biologists estimate that there may be 100,000 to 150,000 black bears in Alaska. Additionally, there are between 6,000 and 7,000 polar bears in the state. In short, Alaska is 'bear country', and the opportunity to view Alaskan bears is a unique and important experience for wildlife viewers. Alaskan residents and non-residents spend more money to view bears than any other wildlife species (Miller and McCollum 1997), and they report deriving greater social benefit from viewing bears than from viewing any other species (Miller et al. 1998).

Researchers from the University of Alaska, Anchorage investigated the spending patterns of brown bear viewers departing from one business in Homer, Alaska (the self-proclaimed 'bear-viewing capital of Alaska'). They found that 545 individual bear viewers spent an average of $2,828 to view bears. By comparison,

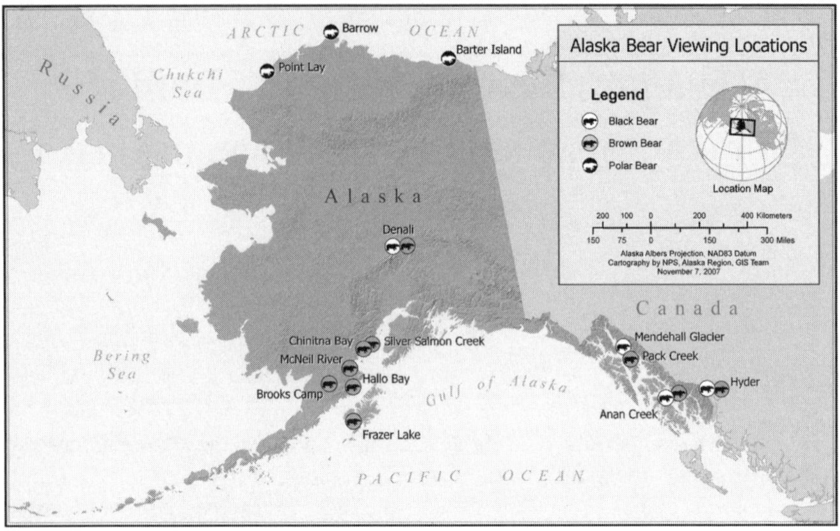

Figure 7.1 Selected bear-viewing areas in Alaska

the typical summer visitor to Alaska spends only about $1,400 (Colt and Dugan 2005).

This chapter explores the bear-viewing experience in Alaska in terms of bear conservation, human safety, and enjoyment. The growing demand for bear-viewing presents a challenge to resource managers as they strive to minimize human impacts on both bears and associated wilderness. Areas that concentrate bears attract the greatest number of viewers. These areas are always important feeding sites, so the potential for negative impacts on bears by viewers is often great. Given the strong growth of the bear-viewing industry in Alaska, human impacts must be monitored and evaluated so managers can control growth and minimize negative impacts. It is essential that agencies develop bear-viewing policies and programmes specific to each bear-viewing area that provide for optimal visitor experiences and safety, while minimizing their impact on bears.

Bear-Viewing Areas on Alaska's Public Lands

Bear-viewing is a recreational pastime and growing industry in Alaska (Aumiller and Matt 1994, Fagen and Fagen 1994, Titus et al. 1994, Wilker and Barnes 1998), and Alaska's public lands provide unparalleled bear-viewing opportunities. In addition to the well-known bear-viewing programmes at the State of Alaska's McNeil River, the US Forest Service's Pack Creek and Anan Creek, and the National Park Service's Katmai and Denali National Parks, bear-viewing opportunities are abundant throughout Alaska's public lands and participation in

Figure 7.2 A sufficiently high prospect of viewing females with offspring in close proximity is an important characteristic of bear-viewing sites for some tourists

Source: T.D. DeBruyn.

viewing programmes is increasing (see Figure. 7.1). For additional information regarding specific bear-viewing sites in Alaska we refer the reader to Sherwonit (1993), Sydeman and Lund (1996), and Stringham (2007).

Along coastal Alaska, brown and black bears congregate throughout the spring and summer in small pockets of salt marsh habitat to feed on nutrient-rich sedges (*Carex* spp.), grasses (*Pucinellia* and *Elymus* spp.), and forbs (*Triglochin* and *Plantago* spp.), as well as adjacent intertidal zone invertebrates, including shellfish and barnacles (Smith and Partridge 2004). Bears also use salt marshes to rear young, as travel corridors, and for courtship and breeding, among other activities. As chum (*Oncorhynchus keta*), pink (*O. gorbuscha*), silver (*O. kisutch*), and red (*O. nerka*) salmon begin entering streams in summer and fall, bears tend to migrate from tidal marshes and move along the coast in search of productive fishing sites. In most coastal habitats where salmon are abundant, brown bear densities can exceed 175/1,000 km^2 (450 bears/1,000 miles^2) (Miller et al. 1997). Sellers et al. (1998) measured 550 brown bears/1,000 km^2 (1,420 bears/1,000 miles^2) in Katmai National Park on the Alaska Peninsula. Many of these foraging areas are readily accessible to humans by boat and small aircraft. Some areas, such as shallow streams and tidal mud flats, which previously limited human use, are now accessed by hovercraft and airboats. As human use in these areas increases so does the potential for negative effects on bears (Wilker and Barnes 1998, Rode et al. 2006a).

Polar bear-viewing is a relatively recent phenomenon, but it is growing in popularity along Alaska's North Slope communities (that is, Barter Island, Barrow, and Point Lay) (Figure 7.1). Inuit whalers harvest bowheads (*Balaena mysticetus*) for subsistence purposes in the fall, and large numbers of polar bears congregate at whale harvest sites near the villages of Kaktovik, Barrow, and Point Lay. A number of private companies offer polar bear-viewing services, bringing visitors to sites where they observe, from vehicles, polar bears feeding on the butchered remains of bowheads.

The demand for bear-viewing exceeds all other ecotourist activities in Alaska because bears are perceived as charismatic, viewing occurs in spectacular wilderness settings, and the likelihood of observing bears is a certainty in a number of key areas (Figure 7.1). Additionally, in these areas there exists a high probability of viewing females with small cubs in close proximity, a favourite photographic subject (Figure 7.2).

Key Characteristics of Successful Bear-Viewing Programmes

Stephen T. Mather, the first Director of the National Park Service, once said that 'Scenery is a hollow enjoyment to the tourist who sets out in the morning after an indigestible breakfast and a fitful night's sleep ...' Along with Mather's advice for viewer satisfaction, we suggest that management agencies consider a number of important elements associated with effective bear-viewing programmes.

Foremost, agencies need a well-defined mission for their bear-viewing programme and managers should have a clearly-articulated operations plan and goals in place prior to initiating bear-viewing at a given site. An over-arching principle that should guide the planning process is that 'bears come first'. Therefore, any action that may adversely affect bears' access to forage resources should be avoided. While bear viewers should be reasonably safe and accommodated; high quality bear-viewing sites should have visitor facilities such as bathrooms and places to escape the weather, this should not be to the detriment of natural patterns of bear activity in the area.

Temporal and spatial predictability of site-usage patterns by bears is required, but physical contact of tourists with bears should be minimized (for example, use of aversive conditioning and translocation). The site must be accessible, but accessibility should not alter bear behaviour nor impact their existing wilderness quality. Viewers' movements should be temporally and spatially predictable for bears (for example, people should travel the same routes at approximately the same times both to and from the site). Nearby alternative viewing sites should be available as this alleviates pressures on bears and affords increased viewer opportunities, whilst reducing viewer angst and boredom.

Table 7.1 Characteristics of successful bear-viewing programmes

Site Characteristics	Management Characteristics
• The site is easily accessible to visitors • Accessibility is not weather dependent • Accessibility does not impact on the visitor experience or bear activity • Bear visitation patterns are predictable • There is a large concentration of bears • Observation points are close to bear activity • Viewing structures assure visitor safety • Alternative viewing areas are available nearby	• Bears are given priority in movement and access to forage resources • Informative interpretation about bears and ecosystem processes is provided • Management affords a quality viewing experience • Staff are well trained, with consequent low turnover • Facilities are managed consistent with expectations • Viewing programmes are optimized for local conditions • Managers are responsive to viewers' expectations and needs • The management of the site involves local communities and incorporates native peoples and cultures

The objectives set by management should be measurable. This requires resource assessment prior to and during programme initiation to ensure that the effects of changes induced by bear-viewing can be detected. For management to be data-driven and responsive, monitoring of bear populations and their environment should occur at each site.

Agencies need to pay qualified personnel a living wage which promotes longevity and local experience with bears. A veteran staff will help institutionalize management, keep it consistent, and over time will ultimately improve the visitor experience and increase safety. The consequent cost of implementing a bear-viewing programme can be substantial and this should be considered carefully prior to implementation.

Guides at agency viewing sites should be agency personnel and they should be certified through bear-viewing guide certification programmes. Both guides and managers should be experienced not only with bear behavior, but with the viewing sites' natural history and with the management procedures at other bear-viewing sites. Conveying information about the ecosystem in which bears are being viewed adds immeasurably to the visitor experience: interpretation is a key element of a successful programme. Significant cultural aspects of the viewing site should also be included in the bear-viewing programme. However, bear-viewing sites should not be over-promoted to help keep the expectations of viewers realistic.

As a final point, angling and bear-viewing do not generally mix. If angling occurs, then bear-viewing should not be promulgated by the management agency. The key characteristics of successful bear-viewing programmes, in terms of both site and management attributes, are summarized in Table 7.1.

Managing Bear-Viewing Areas on Alaska's Public Lands

The growing demand for bear-viewing and associated photographic opportunities presents a challenge to resource managers seeking to accommodate viewers whilst minimizing human impacts. While Alaska's land management agencies strive to provide quality bear-viewing opportunities for the public, they also have a responsibility to maintain healthy bear populations. Areas preferred by bear viewers are nearly always critical feeding sites with a high concentration of bears, and as human presence in these areas increases, so the potential for negative impacts on bears using the feeding sites also increases (Olson et al. 1997, Wilker and Barnes 1998). Bear populations concentrated around feeding sites provide exceptional viewing opportunities for the public, but these sites must also be managed to minimize the impact on bears.

Bear-viewing in Alaska most often occurs where bears aggregate to forage on seasonally plentiful food sources such as salmon in coastal streams or salt marsh vegetation located along the coast. Bear use patterns of salt marsh meadows are dependent on the amount of protein in the forage and forage biomass (Smith and Partridge 2004); seasonal variations in use of salmon streams is dependent on the availability of salmon. The presence of bear viewers can affect bears' use of these foraging sites (Rode et al. 2006a). Across North America bears have been observed avoiding areas of intense human activity (Warner 1987, Gunther 1990, Fagen and Fagen 1994, Smith 2002). Bear-viewing programmes in some areas have led to the displacement of individuals (Warner 1987, Olson et al. 1988), particularly adult males which have lower disturbance thresholds than other bear cohorts (Smith 2002). However, the costs of displacement at the individual and population levels are difficult to assess and have not been well-studied.

It is useful to analyse bears' response to human disturbance within the theoretical framework of anti-predator behaviour. Frid and Dill (2002) noted that animals rarely have perfect information, and generally are expected to maximize population fitness by overestimating, rather than underestimating, risk. Overestimating risk has less severe consequences than its underestimation, the result of which might be death. Because hunters have presented a real threat to many species for millennia, disturbance and predatory stimuli may be indistinguishable from the animal's perspective (Frid and Dill 2002). Thus, habituation to disturbance stimuli is often negligible, and presumably maladaptive. In short, animals simply cannot risk ignoring people totally. However, for some species wildlife viewing may have no impact other than to alter their behaviour (that is, no effect on individual fitness, population, or community). For this reason, some researchers believe that study needs to be directed only toward those species that are threatened or to populations that are declining in order to avoid unnecessarily restricting access to wildlife viewing areas (Gill et al. 2001).

Bears will alter their temporal patterns to utilize those time periods that are consistently and predictably absent of humans (Smith 2002, DeBruyn et al. 2004, Rode et al. 2006a). However, recent research suggests that some segments of bear populations may benefit from bear-viewing via displacement. Nevin and Gilbert (2005a) found that the presence of bear viewers in British Columbia displaced large male bears, thus allowing female bears and their offspring access to otherwise unavailable resources. Rode et al. (2006b) drew similar conclusions from their research at Douglas River, Alaska. Researchers in that study proposed that resource use by female brown bears with offspring was significantly related to the presence of large male bears and that sexual dimorphism and different reproductive strategies explained segregation at foraging sites. Others, however, are guarded about touting this positive effect of bear-viewing when not evaluated in an area-specific context. Tollefson et al. (2005) note that male dominance does not occur at many fishing sites in Alaska where fishing efficiency or intake is insufficient to attract large males; that male displacement will not occur at some well established bear-viewing sites where males are habituated to people and, perhaps due to the prolonged exclusion of hunting, perceive no risk from humans (such as in national parks); and that both sexes of bears will be displaced from salmon streams by people in areas where all bears perceive a major risk from humans, who occasionally kill them for a variety of reasons (some legal, some not). In situations like those mentioned by Tollefson et al. (2005), the positive aspects of male displacement on female bears with offspring are likely to be diminished or non-existent. Moreover, few studies have examined how the effects of human behaviour at viewing sites and the natural vigilance/predation risk behaviour of animals intertwine to affect the viewed animals (Peters and Otis 2005). Absent demographic data studies that rely on behavioural data alone cannot estimate the true costs of behavioural changes to animals caused by wildlife viewing (Griffin et al. 2007).

Harassment and chronic disturbance of wildlife have been shown to be detrimental and to cause decreases in reproduction (Yarmoloy et al. 1988). Bears habituated to the presence of people exhibit fewer signs of stress than do non-habituated bears (Herrero 1989). Habituation of bears to people is believed to be largely responsible for the success of at least one major bear-viewing area in Alaska (Aumiller and Matt 1994). However, Smith et al. (2005) suggest that bear-viewing has been a surprisingly safe activity due to bears' heightened tolerance of conspecifics at resource-rich sites, not due to habituation to humans. Regardless, human behaviour can negatively impact bear aggregations. For example, on Kodiak Island, Alaska, Wilker and Barnes (1998) demonstrated that people participating in structured bear-viewing programmes were less disruptive to bears than was the

2 The potential benefits and risks (costs) of bear-to-human habituation

Benefits Primarily To Humans	Costs (Risks) Primarily To Humans
• Provides highly sought after opportunities for acceptably safe bear-viewing, photography, film-making. • Bear-viewing is a growing industry in North America offering economic benefits to many. • Evidence that habituated brown bears are less likely to threaten or attack hikers or bear viewers. • Habituation and use of roadside habitat could in theory increase the carrying capacity of protected areas for brown bears.	• Not appropriate in all contexts; may conflict with sport fishing and hunting. • Close proximity of habituated bears may encourage ignorant, even illegal acts. • More interactions with habituated bears may increase cumulative odds of injury. • Habituated brown bears, especially sub-adults, have a greater tendency to approach people, and people may respond inappropriately and dangerously. • Habituated bears in roaded areas may encourage traffic jams and serious collision injuries. • It costs money to manage habituated bears, especially if they become food-conditioned.
Benefits Primarily To Bears	**Costs (Risks) Primarily To Bears**
• Habituated bears are better able to access natural foods and other resources that exist near centres of human activity. • Some bears may use presence of humans to avoid encounters with other bears (mothers caching cubs near observation stations). • Habituated bears promote bear viewing which, in turn, may promote bear conservation.	• Habituated bears near roadsides or railways are more likely to be injured or killed. • Habituated bears are more likely to be killed if outside of protected areas. • Habituated bears near roads are more likely to be fed by people or get people's food, and become food-conditioned. • Despite regulations, habituated bears are more likely to be approached by people for better photographs or viewing, resulting in greater risk of human injury and bear harassment or removal.

Source: Herrero et al. 2005.

case with programmes lacking consistency and predictability. Similarly, at Brooks Camp, Alaska, DeBruyn et al. (2004) showed that certain human behaviours displaced bears from trails. However, even habituation to motorized traffic by wildlife is achievable if activity is initially confined to predictable routes and times (Yarmoloy et al. 1988, Smith 2002). Yet, even though bears may habituate to anthropogenic stimuli under certain circumstances, the process that culminates in habituation/tolerance is not well understood. Research suggests that there are threshold levels of stimuli to which wildlife will not habituate. For example, Miller and Gunn (1980) have shown 200 m above ground level (agl) as a statistically significant threshold elevation below which extreme responses occurred more often for Muskox (*Ovibos moschatus*) when harassed by helicopters. Similarly, Wilker and Barnes (1998) noted that fixed-wing aircraft flying below 100 m agl elicited strong responses from bears, and that the incidence of strong response declined sharply at 100 m to 200 m agl, and was negligible at 200 m agl. DeBruyn et al. (2004) raised caution that, while bears may habituate to temporal-spatial use patterns of bear viewers, there may be limits to the amount of displacement bears can tolerate and still efficiently exploit the resource. The authors further noted that curtailment of the disruptive impacts of bear viewers may not be affected through outreach alone, but requires a commitment by the management agency for intensive on-site management to effectively eliminate disruptive viewer behaviour.

Regardless of proximate mechanisms, most bears at viewing areas seem to tolerate humans. An assessment of the potential benefits and risks (costs) to people and bears of habituation (Table 7.2) should be completed for all areas advertised for public bear-viewing (Herrero et al. 2005). Canadian researchers recently applied such a cost/benefit/risk approach to evaluating the suitability of bear-viewing for three different situations in the Yukon Territory and the decision regarding whether or not habituation was desirable appeared clear in each case (S. Herrero, University of Calgary, Canada, pers. comm.).

Effective management of brown bears at bear-viewing sites requires site-specific information regarding bears' responses (changes in behaviour and use of resources) to potential anthropogenic stressors including the effects of aircraft, all-terrain vehicles (ATV), numbers of human visitors, and bear-viewing activities. Managers can use this information to define aircraft landing areas, minimum overflight altitude, placement of bear-viewing facilities, areas and times of access, and other use guidelines/restrictions to mitigate human impacts on bear populations as well as individual bears. Long-term monitoring can then determine whether management actions are effective in achieving desired outcomes and adjustments made through adaptive management.

While some are concerned about the impact of increased wildlife viewing on wildlife and their environment, others are concerned that increased bear-viewing will exclude them from the opportunity to hunt in some areas. Conflicting values regarding bear-viewing and hunting have lead to what some call a 'culture war' between these factions in Alaska. However, almost no empirical data exist to support either side's assertions. The basic premise, that bear-viewing leads to human-habituated bears and, in turn, susceptibility to unfair harvest potential, is unstudied and unproven. Because it is un-quantified to what extent bear-viewing may habituate bears to humans, vis-à-vis other factors, clearly the assertion that habituated bears are easier to hunt than those not associated with bear-viewing is untested. Because some degree of habituation is likely to occur around all bear-viewing sites, hunting near those areas poses an ethical dilemma and is quite controversial. Management objectives must be clearly set prior to the commencement of formal viewing activities. If hunting is desired, viewing programmes should avoid habituation and non-concurrent seasons should be established for both activities. Clearly, the dynamics of bear-viewing (many persons observing the natural behaviour of an aggregation of bears) is unlike those of bear hunting (a few individuals dispersed and secreted on the landscape awaiting a desirable bear to walk into their rifle sights). To date, debates regarding the appropriateness of hunting and viewing the same bear population have been based largely on conjecture, and not the careful research necessary to inform management decision-making. For these reasons, research designed specifically to answer these important questions is long overdue.

Recommendations for Managing Bear-Viewing Areas on Alaska's Public Lands

The dramatic increases in commercial bear-viewing that are taking place in the coastal areas and inland rivers of Alaska have resulted in situations where viewer safety and bear management operates more by default than design. It is imperative that the agencies identify areas where bear-viewing opportunities will be provided and develop policies to address human safety and natural resource concerns. Within agency lands some bear-viewing is structured, with visitors and bears managed by well-defined plans, while some bear-viewing is unstructured, with no over-arching management policies. It is our opinion that land management agencies currently lack a unifying philosophy that guides bear-viewing programmes. While bear conservation should be a fundamental priority at bear-viewing sites, managers seemingly pay most attention to accommodating the wishes of people over and above considering the impacts on bears. For this reason, we feel it is important to define a regional approach to bear-viewing within Alaska. As bear-viewing continues to increase in popularity, agencies will need to prepare for and manage a burgeoning visitor presence. Potential bear-viewing sites need to be evaluated so that each agency can select alternative locations for expanded bear-viewing, rather than wait for unregulated programmes to develop haphazardly. It is essential that the agencies develop a joint regional bear-viewing policy that recognizes a site's carrying capacity and provides for optimal visitor experiences and safety while minimizing impacts to Alaska's bear populations and their habitats. By defining regional bear-viewing policies, Alaska's land management agencies will establish themselves as leaders within the land management agencies of the US, and they can create opportunities to work with private land managers in defining Alaska's bear-viewing policies and nurturing a sustainable programme in Alaska.

Management agencies largely fail to act in unison to reduce the risks to people and bears associated with bear-viewing, due to political interests, jurisdictional turf, and the egos of those involved. Consequently, jointly determined policies and guidelines that have been promulgated have been generalized to achieve agreement. Unfortunately, the 'one-size-fits-all' mentality is not applicable for establishing successful and sustainable bear-viewing programmes. Therefore, we encourage agencies to work co-operatively with the best interests of bears and their viewers in mind. In this manner, bears and people can be better served by programmes that provide the latter with safe, high quality wilderness viewing experiences and the former with tenable circumstances that do not negatively alter their behaviour or diminish their access to vital forage resources.

Site-specific guidelines need to be considered when managing bear-viewing areas (Tollefson et al. 2005). These guidelines should include site-specific restrictions that will alleviate negative impacts on bears. We suggest that site-specific management includes provisions that address the following:

- Adopt management guidelines for visitor access, movement, group size, and the manner in which groups conduct themselves to minimize impacts on all bear cohorts.
- Define designated pathways where people may walk and have access. If left undefined visitors will not, in all likelihood, pick optimal paths in terms of minimizing impacts to bears.
- Prohibit people from directly approaching foraging bears in certain areas (that is, areas 'closed' to humans).
- Restrict the maximum number of people that may be in a single group, the number of groups that may be present at any one time, or the total number of people concurrently present. Without such restriction the quality of the bear-viewing experience will become degraded and bear disturbance will increase.
- Establish restrictions regarding viewer behaviour. Should groups remain tightly bunched or be allowed to spread out? Should people be allowed to approach bears directly in non-restricted areas and, if so, how close? Should bear-viewing groups remain stationary when approached by bears or be advised to move away when the bear gets within some minimum distance? Should groups be required to carry some form of bear deterrent? These are questions that can, and should, be addressed as part of an overall management approach, prior to human use of bear-viewing sites.
- Establish and enforce minimum human-to-bear distance rules, with the caveat that there are circumstances where maintaining a set buffer is impossible. Bear-distance guidance is a very important and complex issue requiring further discussion. One might consider allowing bears to approach as close as they choose, even up to just a few feet from the

observers (Herrero et al. 2005). This is a respectful way of letting the bears determine their comfort level.

- Impose temporal viewing restrictions. Bear-viewing should be restricted to set times of day as well as seasons, which can be determined on a site-specific basis. For example, the National Park Service restricts human access at Brooks Falls from 21.00 to 06.00 hours, leaving these times undisturbed so that bears sensitive to human activity can gain access. Additionally, seasonal viewing restrictions will help reduce conflict with hunting activities and allow bears time at critical food sources (such as salmon runs and coastal meadows) without human disturbance, and allow human activity to be predictable.
- Consider access restrictions. In some areas bears should be left an option to feed without human observation. There is also concern regarding potential impacts from helicopter and fixed-wing overpasses for viewing purposes, as well as jet-boat tours. The potential disturbance of such activities on bears is unknown and requires further investigation. Should such access methods be shown to be disruptive, they should be restricted or modified accordingly.
- Commercial bear-viewing groups should be led by licensed guides who receive training in bear behaviour, natural history, ethics, agency regulations and first aid.
- Training for the above should be provided by an interagency group representative of Alaska's wildlife and land management agencies.

Conclusions

Bears are part of the landscape and should not be considered separate from it. It may be true that humans and bears have co-existed in Alaska for thousands of years, but in recent times the upsurge in human backcountry use raises concerns regarding the future of bears. Bears figure largely in the culture, ancestry, conservation and politics of Alaska. However, it is only relatively recently that bear-viewing has come to the fore in the Great Land. To effectively manage bear-viewing in Alaska the history and role of humans in the ecosystem cannot be ignored. Some bear-viewing areas today may be artifacts of recent management regimes rather that long-term 'natural' situations (Birkedal 1992). If such history is ignored, both management and viewer expectations may prove unrealistic.

Responsible management agencies need to ensure that bear conservation, human safety and the quality of the experience are the priority objectives. Bear-viewing areas should be monitored to ensure that participants are receiving guidance on bear behaviour, acceptable human behaviour, and proper food and waste handling. The impacts on bears of viewing programmes both within and outside public lands are potentially considerable, and bear-viewing must not compound effects currently faced by individual bears and bear populations.

A dearth of research exists on hunting habituated bear populations and this needs to be remedied. For example, the question 'is habituation site-specific and not transferable from site to site?' begs an answer. A better understanding of how viewing activities affect habituation and how habituation might diminish bears' response to hunting is urgently needed. Because most of Alaska is open to bear hunting, we recommend that bear-viewing and bear hunting are not undertaken in the same location until more is understood about the effects each has on the other.

Ecotourism is a rapidly expanding industry in Alaska and is often touted as an economic incentive and stimulus for wildlife conservation. However, we find the conservation implication for bears in Alaska equivocal at present. Bear-viewing affects bears, sometimes in unexpected ways (Nevin and Gilbert 2005b). Because populations are comprised of individuals, and because individuals, particularly in highly intelligent species like bears, have variable behaviour, predicting the population-level consequences of wildlife viewing based on mean responses can be misleading and therefore ultimately uninformative. What will prove unifying to our understanding will be to study the effects of bear-viewing on the individual bears in a population as it regards their behaviour, habitat use and energetics. Because responses change over time, and from location to location, studies need to be long-term. Such studies can be carried out, but they need to address the site-specific considerations, interagency cooperation and industry involvement that we have outlined. Ultimately, proper management makes not only good conservation sense, but good business sense too.

Acknowledgements

We thank Jennifer Hill and Tim Gale for helpful review and editorial comments. We are grateful for a kindly review of an early manuscript by John Schoen, Senior Scientist, Audubon Alaska. We appreciate the assistance of Jeff Bennett, NPS Alaska for creating Figure 7.1.

References

Aumiller, L. and Matt, C. (1994) Management of McNeil River State Game Sanctuary for viewing of brown bears. *International Conference on Bear Research and Management*, 9, 51–61.

Birkedal, T. (1992) *Ancient Hunters in the Alaskan Wilderness: Human Predators and their Role and Effect on Wildlife Populations and the Implications for Resource Management*. George Wright Society Meeting Presentation.

Colt, S. and Dugan, D. (2005) *Spending Patterns of Selected Alaska Bear Viewers: Preliminary Results from a Survey*. Institute of Social and Economic Research, University of Alaska, Anchorage. Version 17.

DeBruyn, T.D., Smith, T.S., Proffitt, K., Partridge S. and Drummer T.D. (2004) Brown bear response to elevated viewing structures at Brooks River, Alaska. *Wildlife Society Bulletin*, 32, 1132–1140.

Fagen, J. and Fagen, R. (1994) Interactions between wildlife viewers and habituated brown bears, 1987–1992. *Natural Areas Journal*, 14, 159–164.

Frid, A. and Dill, L. (2002) Human-caused Disturbance Stimuli as a Form of Predation Risk. *Conservation Ecology* [online], 6(1): 11. Available at: http://www.consecol.org/vol6/iss1/art11/ [accessed: 10 October 2003].

Gill, J.A., Norris, K. and Sutherland, W.J. (2001) Why behavioural responses may not reflect the population consequence of human disturbance. *Biological Conservation*, 97, 265–268.

Grambsch, A.E. and Fisher, W.L. (1989) History of development of the National Survey of Fishing, Hunting, and Wildlife-Associated Recreation. *Wildlife Society Bulletin*, 17, 538–543.

Griffin, S.C., Valois, T., Taper, M.L. and Mills, L.S. (2007) Effects of tourists on behavior and demography of Olympic Marmots. *Conservation Biology*, 21, 1070–1081.

Gunther, K.A. (1990) Visitor impact on grizzly bear activity in Pelican Valley, Yellowstone National Park. *International Conference on Bear Research and Management*, 8, 73–78.

Herrero, S. (1989) The role of learning in some fatal grizzly bear attacks on people. In M. Bromley (ed.) *Bear–people Conflicts: Proceedings of a Symposium on Management Strategies*. Northwest Territories Department of Renewable Resources, 6–10 April 1987, Yellowknife, Canada, pp. 9–14.

Herrero, S., Smith, T.S., DeBruyn, T.D., Gunther, K. and Matt, C.A. (2005) From the Field: Brown bear habituation to people—safety, risks, and benefits. *Wildlife Society Bulletin*, 33, 362–373.

Miller, F.L. and Gunn, A. (1980) Behavioral responses of muskox herds to simulation of cargo slinging by helicopter, Northwest Territories. *Canadian Field Naturalist*, 94, 52–60.

Miller, S.D. and Schoen J. (1999) Status and management of the brown bear in Alaska. In C. Servheen, S. Herrero and B. Peyton (eds.) *Bears: Status Survey and Conservation Action Plan*. IUCN/Species Survival Commission. Gland, Switzerland and Cambridge, pp. 40–46.

Miller, S.D., White, G.C., Sellers, R.A., Reynolds, H.V., Schoen, J.W., Titus, K., Barnes, V.G. Jr., Smith, R.B., Nelson, R.R., Ballard, W.B. and Schwartz, C.C. (1997) Brown and black bear density in Alaska using radio-telemetry and replicated mark-resight techniques. *Wildlife Monograph*, 133, 1–55.

Miller, S.M. and McCollum, D. (1997) *Alaska Nonresident Voters: Their Attitudes Towards Wildlife and Wildlife Related Trip Characteristics and Economics*. Alaska Department of Fish and Game, Juneau, Alaska.

Miller, S.M., Miller, S.D. and McCollum, D. (1998) Attitudes toward and relative value of Alaskan brown and black bears to resident voters, resident hunters, and nonresident hunters. *URSUS*, 10, 357–376.

Nevin, O.T. and Gilbert, B.K. (2005a) Measuring the cost of risk avoidance in brown bears: further evidence of positive impacts of ecotourism. *Biological Conservation*, 123, 453–460.

Nevin, O.T. and Gilbert, B.K. (2005b) Perceived risk, displacement and refuging in brown bears: positive impacts of ecotourism? *Biological Conservation*, 121, 611–622.

Olson, T., Gilbert, B. and Fitkin, S. (1988) *Brown Bear Behavior and Human Activity at Salmon Streams in Katmai National Park, Alaska*. Final report for USDI, contract IA 9700–7–8028.

Olson, T., Gilbert, B. and Squibb, R. (1997) The effects of increasing human activity on brown bear use of an Alaskan river. *Biological Conservation*, 52, 95–99.

Peters, K.A. and Otis, D.L. (2005) Using the risk-disturbance hypothesis to assess the relative effects of human disturbance and predation risk on foraging American Oystercatchers. *The Condor*, 107, 716–725.

Rode, K.D., Farley, S.D. and Robbins, C.T. (2006a) Behavioral responses of brown bears mediate nutritional effects of experimentally introduced tourism. *Biological Conservation*, 133, 70–80.

Rode, K.D., Farley, S.D. and Robbins, C.T. (2006b) Sexual dimorphism, reproductive strategy, and human activities determine resource use by brown bears. *Ecology*, 87, 2636–2646.

Sellers, R.A., Miller, S., Smith, T.S. and Potts R. (1998) *Population Dynamics of a Naturally-Regulated Brown Bear Population on the coast of Katmai National Park*. Project Final Report, Alaska Department of Fish and Game, Juneau, Alaska.

Sherwonit, B. (1993) Bear-viewing. *Alaska Geographic, Alaska Bears*, 20, 57–69.

Smith, T.S. (2002) The effects of human activity on brown bear use of the Kulik River, Alaska. *URSUS*, 13, 153–163.

Smith, T.S. and Partridge, S. (2004) Dynamics of intertidal foraging by coastal brown bears in Southwestern Alaska. *The Journal of Wildlife Management*, 68, 233–240.

Smith, T.S., Herrero, S. and DeBruyn, T.D. (2005) Alaskan brown bears, humans, and habituation. *URSUS*, 16, 1–10.

Stringham, S. (2007) *Bear Viewing in Alaska*. The Globe Pequot Press, Guilford, CT.

Sydeman, M. and Lund, A. (1996) *Alaska Wildlife Viewing Guide*. Falcon Press Publishing Company, Helena and Billings, MT.

Titus, K., Trent, J., Aumiller, L., Westlund, J. and Sigman, M. (1994) Managing brown bears as both game and nongame: past experience and future prospects. *Transactions of the 59th North American Wildlife and Natural Resource Conference*, pp. 353–362.

Tollefson, T.N., Matt, C., Meehan, J. and Robbins, C.T. (2005) Quantifying spatiotemporal overlap of Alaskan brown bears and people. *Journal of Wildlife Management*, 69, 810–817.

US Fish and Wildlife Service National Survey of Fishing, Hunting, and Wildlife-Associated Recreation (2007) [online] Available at: http://federalasst.fws.gov/surveys/surveys.html [accessed: 1 August 2007].

Warner, S.H. (1987) Visitor impact on brown bears, Admiralty Island, Alaska. *International Conference on Bear Research and Management*, 7, 377–382.

Wilker, G. and Barnes, V. (1998) Responses of brown bears to human activities at O'Malley River, Kodiak Island, Alaska. *URSUS*, 10, 557–561.

Yarmoloy, C., Bayer, M. and Geist, V. (1988) Behavior responses and reproduction of mule deer, *Odocoileus hemionus*, following experimental harassment with an all-terrain Vehicle. *Canadian Field Naturalist*, 102, 425–429.

Chapter 8

Pugmarks, Tyre-Tracks and Footprint Actual and Potential Roles of Ecotourism in Protecting Wild Tigers in India

Simon Evans

Introduction

> A large pugmark is etched in the soft earth before us belonging, we are reliably informed, to one of the more mature resident tigers of the park. We look on in awe. The tyre-tracks of our jeep lead to the spot and, after carefully climbing down from the retractable foot-plate, our guide deposits a similarly fresh but by no means as impressive footprint beside that of the tiger. Our quarry is close, we are told; it could be watching us this very moment from the nearby cover. We load our cameras and carefully crawl along in first gear, another vehicle immediately behind us, a choking cloud of dust following on in our wake. The hunt is on.

The words above are abstracted from my field notebook, covering a study visit to Ranthambhore National Park in 2001. Located within the northern state of Rajasthan, Ranthambhore is widely viewed as one of the 'jewels in the crown' of India's network of reserves comprising Project Tiger, a national commitment to wildlife conservation. It is also one of the most heavily visited tiger reserves in India and enjoys a reputation as one of the best locations to view tigers in the entire subcontinent, making it an important tourism hotspot (Sippy and Kapoor 2001).

In 2001, there were signs that the plight of India's tigers was far from assured, with various concerns being raised by conservationists across tiger range countries as to the precarious nature of their continuing existence in the wild. Since that time, these concerns have been magnified, culminating in what has been labelled the 'third tiger crisis' as news of the local extinction of tigers in another of India's premier reserves, Sariska, was confirmed early in 2005. A comprehensive review of the entire Project Tiger estate was subsequently undertaken and results of this process suggested that, far from being an isolated case, tiger numbers across the range were in serious decline, bringing into question the validity and accuracy of the annual tiger census.

While the main cause of this decline has been attributed to large-scale poaching to service the lucrative, yet illegal, trade in wildlife body parts for use in traditional

Chinese medicine (TCM), in reality the problem has proved to be far more complex in nature and, as a corollary, the range of potential solutions are increasingly difficult to reconcile. At the heart of this debate has been an ongoing critique of established conservation applications, described by some as constituting a Western-driven global ecology that has little relevance to the socio-political climate of modern day India or, at a more micro level, to the needs of impoverished, livelihood-insecure communities living within and around park boundaries. To date, the issue of tourism has assumed a somewhat peripheral role in this debate, due in large part to a growing scepticism on the part of the conservation community as to the actual ability of the tourism industry to convey the full range of benefits promised within the expanded set of objectives that constitute ecotourism.

As will be discussed later in this chapter, the perceived performance of ecotourism in tiger reserves has been compromised by entry into the market of a number of tourism enterprises cynically employing the 'eco-' prefix as a marketing label without ever having to justify their performance in relation to conservation or the pursuit of long-term sustainability. It is crucial that some form of enterprise validation is undertaken in this regard, and that it is supported by a robust and transparent accounting procedure that communicates the merits and limitations of the tourism sector to the discerning ecotourism market, enabling visitors to patronize businesses that contribute most effectively to tiger conservation. It has been evident from interviews with tiger tourists that, whilst visitors may believe that a share of the profits deriving from their travel are invested back into conservation, evidence to validate such claims is often at best anecdotal, and in many cases totally unsubstantiated. It remains the contention of this chapter that the fractured relationship between the conservation and tourism sectors represents a fundamental flaw and, indeed, a major oversight in the development of coordinated strategies designed to manage tigers and their habitats effectively into the future. As will be illustrated throughout the following pages, it is only through the closer integration and reconciliation of these two important disciplines that managers can ensure that the coincidence of pugmark, tyre-track and footprint are attributed to tourists rather than poachers, enabling the tourism sector to play a central and supportive role in sustaining wild tiger populations over the long-term. To paraphrase Milton (2002), the task is to ensure that tigers are, in the first instance, worth more alive than dead and, secondly, that they are valued more highly in the wild than in domesticated situations.

This chapter, therefore, considers the actual and potential roles of ecotourism in protecting vulnerable tiger populations in India. It questions the tourist-centred focus of much current tourism research and argues that, in order to pursue sustainable tourism, the context of visitation needs to be broadened to encompass all stakeholder values. Rather than adopting an isolationist approach that views ecotourism as a panacea to tiger decline, the paper goes on to suggest that an integrated strategy that views tourism development as a component of sustainable development, rather than as a sustainable option in its own right, is vital if the current trajectory of tiger numbers is to be challenged. The chapter is structured in the following way. First, it reviews the current state of tiger numbers and the range of efforts to protect them in the wild. This section includes a discussion of the contested nature of conservation efforts, identifying conflicting platforms of management action and choice. This is followed by consideration of ecotourism, its potentials and limitations, and the current orientation of tiger tourism in India. Finally, and most importantly, the chapter seeks to locate a remodelled ecotourism within the wider context of efforts to protect and expand tiger numbers both within and beyond the existing protected area network.

Tigers in India: A Contemporary Review

> Wild tigers are in a precarious state. Our best approximation concludes that tiger habitats throughout India, Indochina and Southeast Asia are now 40 per cent less than what we estimated in 2005. As the Economic Tigers of Asia leap onto the world stage, wild tiger populations in those countries are in steep decline; today, tigers occupy a mere 7 per cent of their historical range and the threats are mounting, rather than diminishing (Dinerstein et al. 2006: 3).

Research into current tiger numbers across traditional ranges has established a considerable gulf between official estimates and reality, highlighting the ineffectiveness of current census procedures (see, for example, Karanth 2006). This situation received international exposure in 2005 following an announcement by the Wildlife Institute of India (WII) that, contrary to census returns, there were in fact no tigers left in the Sariska tiger reserve in Rajasthan and that a parallel decline was evident in other Project Tiger reserves comprising the national estate. Whilst the primary cause of this problem was firmly attributed to poaching activity to serve the burgeoning TCM market, the underlying pressures can be seen to be far more complex. The debates underpinning poaching for TCM are well rehearsed and lie beyond the immediate focus of this chapter. Those interested in more information on this subject are directed towards the works of Meacham (1997), EIA (2004) and Thapar (2006). The main debate here, however, is less concerned with the final product as with the dynamics of the poaching enterprise, not least the reasons behind the support and collusion of local communities with poachers for short-term financial gain.

India is a country of over one billion people, characterized by a significant disparity between rich and poor. Luce (2006) estimates that approximately 750 million people continue to live in India's 680,000 villages, many of which are geographically and socio-culturally remote from major economic centres. A high proportion of the communities contained within these peripheral areas are characterized as belonging to the rural poor, suffering in many cases from serious livelihood insecurities. Communities of this nature, or at least segments of them, rely heavily upon natural resources to satisfy their basic needs, while others may undertake subsistence agriculture operations or livestock rearing to earn a living for themselves and their families. Karanth and Gopal (2005) estimate that, in addition to its three-quarters of a million people, rural India also supports around 450 million head of livestock. This can have major implications for tiger conservation, not only in terms of competing demands for resources but also due to tigers killing livestock, prompting retribution from villagers through revenge killings (see, for example, Project Tiger 2001). Villages of this type coincide with tiger reserves across India, bringing vulnerable human populations into direct conflict with the wildlife that is being protected within reserve boundaries. Greenough (2003) reviews the problems afflicting the Project Tiger initiative, introducing what he refers to as a series of 'bio-ironies' diminishing efforts to protect declining tiger populations, the majority of which surround not only human-wildlife conflicts but also human–human conflicts, particularly in relation to inequality in the treatment of different societal sectors. Perhaps the most important of these bio-ironies, in the context of this chapter, is the fact that 'actual success in increasing tiger numbers leads directly to a public rage to kill them' (Greenough 2003: 178). Indeed, this is a feature of much of the current dialogue within the effective conservation of mammalian diversity (see, for example, Entwistle and Dunstone 2000, McCarthy 2004).

Seidensticker et al. (1999) link the plight of tigers inextricably to the fact that their habitats exist primarily within human-dominated environments and, in an attempt to alleviate the ensuing conflicts, the forced expulsion and relocation of villages represents a cornerstone of preferred conservation policy, an approach that typifies what Lewis (2004) refers to as 'a global ecology'. The growing tension at this global level, where the values of affluent societies are elevated at the expense of poverty-stricken local communities, lies at the heart of efforts to restructure conservation efforts in line with intra-generational equity considerations (see, for example, Mulder and Coppolillo 2005). This debate is consistent with what Adams and Mulligan (2003) refer to as a pressing need to decolonize nature. This chapter focuses upon this theme, investigating the extent to which ecotourism can play a central role in challenging the negative values of locals through the provision of long-term economic incentives to protect threatened species. Currently, this is not proving to be successful, as evidenced by the extinction of tigers in the Sariska reserve.

The Sariska crisis fuelled the concerns of the international conservation movement that, unless drastic action was taken, the lifespan of tigers in the wild would be unlikely to exceed another fifteen years (see, for example, Eccleston 2007). This situation prompted a major review of tiger conservation in India, opening up a long-standing tension between advocates of two competing conservation paradigms; conservation biology and social ecology. Conservation biology broadly describes the dominant Western-driven platform whereby nature is protected within a series of predominantly 'peopleless' environments, generally referred to as national parks or protected areas. Sarkar (2001: 37) explains the main justification of this approach as arising from 'the fact that many extinctions are known to have been caused by [consumptive] human activities' and removal of these damaging influences enables natural resource managers to concentrate their limited funds specifically upon nature conservation. Terborgh (1999) lends his support to this approach, adding that the presence of ecotourists visiting these sites is, nevertheless, acceptable, desirable and in no way conflicting with the primary objectives of protection. This is due largely to the increased income generated from tourist visitation. But, as Adams (2006) has argued, the ability of protected areas to convey benefits upon all sectors of society, whilst at the same time conserving non-human nature, has proven to be a difficult ideal to achieve. Oppressive poverty contextualized within majestic wilderness remains an inescapable fact of modern conservation.

The inception of Project Tiger was based broadly upon this model, with the establishment of reserves (supported by policies to remove and relocate human communities from inside protected area boundaries), attracting support for tiger conservation through a combination of educative efforts, ecotourism receipts, eco-development and joint forest management measures (see, for example, Datta 2000). Under this regime many communities were forced to surrender their traditional rights to graze cattle, collect firewood and fell trees (Mullan and Marvin 1999), with any return to the reserves treated as illegal incursions. This practice had far-ranging implications for many of India's landless, livelihood-insecure rural communities who found alternative sources of income or subsistence increasingly difficult to secure. This focus upon the exclusion of humans from nature represents a major component of the critique targeting conservation biology.

Perhaps the most vocal opponent of this Western-driven conservation model in India is Guha (1989: 75), who argued that this approach can be 'positively harmful when applied to the Third World' and that:

> the setting aside of wilderness areas [in this way] has resulted in a direct transfer of resources from the poor to the rich. Thus, Project Tiger … sharply posits the interests of the tiger against those of poor peasants living in and around the reserves. The designation of tiger reserves was made possible only by the physical displacement of existing villages and their inhabitants; their management requires the continuing exclusion of peasants and livestock … transplant[ing] the American system of national parks onto Indian soil.

Within this article, Guha advocates the need for a distinctly Indian form of social ecology as a viable alternative to conservation biology, one that explicitly involves local people and rewards them for the continuing presence of wildlife within and around their immediate neighbourhoods. One way of pursuing this is by giving villagers a greater stake in the financial benefits accruing from tourism activity which is, as discussed later in the chapter, wholly consistent with the theoretical foundations and objectives of the ecotourism sector. What is clear, however, is that this is at present not being achieved and that, unless such problems are addressed, the role of ecotourism is set to remain a peripheral influence in efforts to protect tigers from extinction in the wild.

Tiger Tourism: The Present

> I stare straight ahead. Sixty or seventy yards away, something is standing in the middle of the track. I lift up my binoculars to take a look. But with the jeep bouncing like a jackhammer over the ruts and rills, there is no way to tell what it is. It might be a tiger. It might be a Great Dane. It might be a cardboard box. At twenty yards the animal, whatever it is, moves off to the left. Anger flashes through me … (Ives 1996: 22).

This brief passage tells us much about the problems inherent in the current model of tiger tourism, not least the difficulties imposed by landscape conditions in terms of viewing and photographic limitations. Tiger reserves are characterized by an abundance of vegetative cover, vital for the stalking of prey. This cover, however, also helps to shield tigers from the gaze of tourists. Singh (1973), for example, describes the tiger as one of the most difficult subjects to photograph satisfactorily because of its nocturnal and cautious habits. Such limited viewing opportunities can be seen as existing in sharp contrast to growing demands from wildlife tourists for close and intimate encounters with charismatic species (see, for example, Bulbeck 2005). In recognition of this inadequacy, there is a danger that wildlife becomes manipulated in order to provide guaranteed sightings for tourists through, for example, the use of prey species and even domestic livestock as bait. Singh (1973: 193) goes on to describe what he calls 'tiger training', where the regular tethering of bait in a single location conditions tigers to visit a site habitually, enabling guides to establish viewing points where visitors can enjoy an uninhibited view of the predators. However, it is not only this form of direct intervention and manipulation of wildlife that leads to the habituation of (and thus opportunities to view) tigers, for research suggests that the regular presence of tourists at a site can also indirectly alter their behaviour.

Mills (2004: 141) explains that, in the Ranthambhore and Bandhavgarh reserves, tigers have clearly become 'habituated to vehicles, and will stalk and even kill prey close to [tourist] jeeps'. Furthermore, he argues, in areas where poaching pressures are contained, tigers have begun to move about more freely during daylight hours, further expanding potential viewing opportunities for tourists. Mills (2004) does, however, provide a caveat to this observation, recognizing that during the poaching crisis in Ranthambhore in the early 1990s, tigers were rarely seen, suggesting that as tiger numbers contract so they tend to return to their nocturnal habits, limiting potential sightings further. Ward and Ward (2000: 161), whilst recognizing that tourism has a role to play in the conservation of tigers, voice concern as to the growing scale of operations, pointing to the presence of 'long lines of jeeps and Land Rovers [that] now wind through the forest, morning and evening, scattering herds of deer as their eager passengers lean out to shout at one another, 'Have you seen the tiger? Have you seen the tiger?'

In Kanha National Park, in Madhya Pradesh, reserve managers have introduced what they label the 'Tiger Show', in an attempt to maximize tourist–tiger interactions. This practice is undertaken on elephant-back and entails park mahouts tracking tigers early in the morning, before the tourist traffic arrives, until they find a co-operative tiger, 'namely one that is content to sit for a considerable time' (Moulton and Hulsey 2002: 26). They then radio back to headquarters, where waiting visitors are ferried by jeep to a nearby access point and transported by the mahouts to the location of the encounter. These examples illustrate the growing pressure upon reserve managers and tourism providers to manufacture tiger encounters, an issue that is consistent with a growing demand to provide a spectacle as the visual culture of tourism becomes ever more important and tourists are, paradoxically, seeking authenticity as long as this entails sufficient visual stimulation via physical proximity to charismatic wildlife.

Returning to my own visit to Ranthambhore in 2001, the ramifications of this trend can be illustrated through two distinct experiences. The first journey into the park was undertaken in the company of a middle-aged couple from the UK who were extending their vacation to Goa with a brief visit to the 'natural wonders of the north'. Unfortunately, on this visit the anticipated tiger sighting failed to materialize, leaving my companions disappointed and disillusioned. Despite numerous other sightings of what they considered to be 'lesser beasts', the couple remained agitated and, upon their return to the hotel, proceeded to seek recompense for the lack of tigers on view, brandishing a glossy brochure advertising their particular 'tiger tour' complete with an extreme close-up photograph of a tiger on the cover. Returning to the field notebook for that particular visit, I quote: 'the hotel is pictured in the brochure and if we hadn't ever got to see that then we'd be entitled to a refund, so [now pointing at the tiger on the cover] why not for this?' This experience raises an important issue relating to the need for responsible marketing in helping to control tourist expectations.

It is important in this context that visitors to reserves are aware that actually viewing a tiger is by no means guaranteed. It is, after all, a wild and unpredictable animal and if tourists are indeed seeking authentic experiences, then to not see a tiger is perhaps the most authentic visit of all. A sign in Bandhavgarh, just before visitors exit the reserve, perhaps illustrates this point most effectively. Employing a picture of a tiger with a speech balloon coming out of its mouth, the text reads 'perhaps you may not have seen me, but please don't be disappointed. I have seen you'. Moulton and Hulsey (2002: 97) pick up on this theme in connection with the Kanha-based tiger show. 'The fact that the view of a tiger that the show affords is somewhat artificial scarcely seems to concern most tourists, who may be subliminally aware that the entire experience of a national park is less than definitively "natural"'. But while the show has been 'vehemently criticized by purists on a variety of grounds', it does provide guests with their best chance to see and to even photograph a tiger at close quarters. As Dobbie (2004: 120) goes on to argue '… some purists feel [the tiger show] is in some way prostituting wildlife viewing. What nonsense … seeing a tiger at close range from an elephant is one of the most visually splendid experiences one can hope to have …[it] is something to be valued and will always stay in one's visual … memory'. This assertion illustrates how difficult a task it is to control expectations.

My second experience from Ranthambhore in 2001 coincided with the successful sighting of two tigers. This visit was undertaken in the company of three tourists from the USA, all of whom described their experience as 'life-changing' and 'exhilarating', very much in the vein of the comments from Dobbie (2004) included above. This positive experience, however, contributed little to the cause of conservation, but rather to our Mumbai-raised driver who gratefully counted out the generous gratuities passed his way in recognition of his apparent skill in locating viewable tigers. This is another problem associated with the current model of tiger tourism; aside from the entry fees paid at the gate, all other tourist receipts tend to leak from the reserves, either directly to the drivers or back to the distant hotels where the guests stay. Even in the case of the gate receipts, little remains in the parks as they are payable to the state exchequer from which reserve managers have to apply for, and justify, ongoing funding. Padel (2005: 45) points out that similar problems were encountered in Kanha, not least in terms of the inability of local communities to involve themselves in the tourism industry. Here, she saw no locals benefiting from tourism: 'My hotel's owner lived in Delhi and we drove to the reserve past hopelessly poor people who only ate rice'. Perhaps this inability to fully integrate tourism into a broader management response to tiger decline can be explained by Hannam's (2005: 170) assertion that 'the conservationist impulse has largely taken over from any interest in developing tourism. Some eco-development projects have been conceived and implemented … however, the main objective is to conserve biodiversity'. Hannam goes on to argue that this approach does not necessarily encompass tourism development and, as a consequence, this potentially valuable sector continues to operate in isolation of conservation efforts.

Justin Matthews, the Director of the tourism company Discovery Initiatives, puts this issue into context. He claims that:

> Tens of millions of pounds, dollars and rupees are spent every year by tourists in India to experience the extraordinary wildlife that the country has to offer … Sadly, an extraordinarily small percentage of this figure ever reaches the end product – India's wildlife and the local people that live in the vicinity of its influence. Changing this percentage, even by a small fraction, would potentially release huge amounts of money for tigers, forest and wildlife conservation (Matthews 2000).

Moreover, if a proportion of these tourism-generated proceeds are channelled towards local communities, then this will help to close the gap between the values associated with living and dead tigers and may help to erode the support base for poachers that currently exists within the villages. This point is central to the findings of the report of the Tiger Task Force, which recognizes that when used correctly 'tourism can be a potentially viable economic activity [but] … if managed badly, can be a potentially devastating activity for ecology and people'

(Project Tiger 2005: 134). Matthews (2005) follows up on this issue, quoting the Director of Ranthambhore, Govind Sagar Bhardwaj, who claimed that it was the 'big hotels, supported by influential politicians, [that] have a short-term view and keep demanding access for more safari vehicles to ferry growing numbers of tourists to see the tigers'. Mathews adds that currently there is a carrying capacity ceiling of 256 visitors allowed into the reserve at any one time but, with a bed-space capacity of over 1,000 guests in the area, pressures for access continue to grow at an alarming rate.

The presence of so many bed-spaces in proximity to Ranthambhore contributes a direct tourism turnover estimated at around 220 million rupees (£2.8 million) (Project Tiger 2005). Wealthy tourists may spend 30,000 rupees (£40 per night) to stay in the luxurious Aman-i-Khas resort, more than a year's income for many rural Indians and yet, as Foster (2005) asserts, very little of this income is invested back into the park or its people. Moreover, as tourists are ferried to the reserve gates, few if any of the vehicles ever stop en route to interact either socially or economically with local people who enjoy little in the way of water, schools or medical facilities. It becomes clear, therefore, that whilst tiger tourism represents a powerful tool with which to contribute to wider conservation efforts, at present this is not being realized. Perhaps the key question here is whether this form of tourism can even be described as ecotourism or whether it is merely being labelled as such due to its general focus upon wildlife?

Tiger Ecotourism: Remodelling the Product

The discussion above provides some explanation as to why the conservation lobby remains cautious in promoting ecotourism as a key component of protected area policy. Such scepticism, however, is not confined to conservationists alone, but also pervades debates from within the tourism sector itself. Goodwin (2002), for example, challenges the continuing validity of the ecotourism label in light of its increasing application to all types of natural area visitation without necessarily satisfying the multiple objectives that theoretically distinguish this sector from wider tourism markets. Here, it is argued, irresponsible operators and service providers can utilize the 'eco-' prefix as a marketing tool, designed primarily to maximize market opportunities rather than pursue sustainable outcomes. Ceballos-Lascurain (2002), while recognizing certain inadequacies associated with ecotourism in practice, counters this argument by stressing the important contribution that it can play in local and regional development, arguing that we should be less concerned with 'eco-purism' than ecotourism. A problem persists, however, in the range of conflicting definitions attached to the sector (see Fennell 2001), creating difficulties in assessing what ecotourism actually is and how it should be measured. Depending on the definition employed, the economic value of ecotourism generates anything from US$30 billion (Honey 1999) to US$1.2 trillion (Ceballos-Lascurain 1996).

This problem has major ramifications for ecotourism and its acceptance as responsible tourism, as illustrated within the earlier section on the current tiger tourism model as practiced in India. To borrow a quote from Primo Levi (1989: 158), 'to give a name to a thing is gratifying … but it is also dangerous. The danger consists in one's becoming convinced that all is taken care of and that once named the phenomenon has also been explained'. This argument has direct relevance to ecotourism and its relationship to conservation, activities that are currently viewed as divergent. Indeed, perhaps the main convergence between the two sectors is the fact that both tend to exclude local communities from their operations, a process that conflicts with the main contention of this chapter. What has emerged as a consequence of this approach is a view of people as antithetical to species conservation, hosts and visitors alike. Also, in a spatial sense, such practices have culminated in the creation of 'bubbles' in which the activities of both sectors have become focused. In a conservation sense, these bubbles coincide with protected area boundaries. From a tourism perspective, ecotourists are accommodated within sanitized spaces that manage visitor movements and insulate them from external influences (Carrier and Macleod 2005). These spaces include not only the tiger reserves themselves, but also distant hotels from which vehicular transfers to the 'attraction' are necessary. These transit zones coincide with relocated villages, and while this offers theoretical opportunities for local access to tourism markets seldom, if ever, does tourist transportation stop en route. It is viewed as imperative that these restrictive bubbles are enlarged to take account of local aspirations, and that local people benefit financially from the existence of tigers within nearby parks and reserves.

This task requires greater integration of conservation and tourism-related interests and strategies. To achieve this, conservationists need to recognize the potentials of ecotourism and to devise strategies that enable them to circulate the generated receipts more widely. At the same time, ecotourism providers need to validate their claims towards responsible activity, regulating and controlling visitation, and orienting it more closely towards long-term conservation priorities. To achieve this, both the conservation and ecotourism sectors need to broaden their common linkages and networks, entering not only into meaningful dialogue but, crucially, co-operative action.

Finally, another important issue for ecotourism providers surrounds the need to control the aspirations of tiger tourists rather than implying guaranteed viewing opportunities. This is not, however, a straight-forward task, particularly in light of what Bulbeck (2005: 151) refers to as the 'Attenborough effect', whereby the 'up-close, intimate' imagery employed within television documentaries provides a kind of benchmark for what visitors expect to see in the wild. This trend is compounded by marketing literature that tends to hint towards an increasing intimacy between visitors and tigers. It is important that guidelines and sector-specific regulations target this issue, emphasizing the less tangible atmospheric and experiential values attached to a visit to tiger reserves, regardless of whether tigers are seen or not. There are some encouraging signs at present that strides are being made in this direction, particularly under the aegis of the Tour Operators for Tigers (TOFT) initiative, a collaborative effort between representatives of the tourism (Discovery Initiatives) and conservation (Global Tiger Patrol) movements. This represents an important starting point, but is by no means an end in its own right. If tigers are to remain in the wild, such partnerships need to be expanded further, with ecotourism and conservation at the forefront of efforts to ensure that pugmarks, tyre-tracks and footprints continue to coincide on the dusty roads of India's network of tiger reserves.

References

Adams, J. (2006) *The Future of the Wild: Radical Conservation for a Crowded World*. Beacon Press, Boston.

Adams, B. and Mulligan, M. (2003) *Decolonising Nature: Strategies for Conservation in a Post-Colonial Era*. Earthscan, London.

Bulbeck, C. (2005) *Facing the Wild: Ecotourism, Conservation and Animal Encounters*. Earthscan, London.

Carrier, J. and McLeod, D. (2005) Bursting the bubble: The socio-cultural context of ecotourism. *The Journal of the Royal Anthropological Institute*, 11, 315–334.

Ceballos-Lascurain, H. (1996) *Tourism, Ecotourism and Protected Areas*. IUCN, Gland.

Ceballos-Lascurain, H. (2002) Preface. In M. Luck and T. Kirstges (eds) *Global Ecotourism Policies and Case Studies: Perspectives and Constraints*. Channel View, Clevedon.

Datta, S. (2000) A revolutionary conservator blazes a new trail in Srisailam Tiger Reserve. *Forests Trees and People Newsletter*, No. 42.

Dinerstein, E. Louks, C. Heydlauff, A. Wikramanayake, E. Bryja, G. Forrest, J. Ginsberg, J. Klenzendorf, S. Leimgruber, P. O'Brien, T. Sanderson, E. Seidensticker, J. and Songer, M. (2006) *Setting Priorities for the Conservation and Recovery of Wild Tigers: 2005–2015*. Smithsonian, Washington DC.

Dobbie, A. (2004) *India: The Tiger's Roar*. Melrose, Ely.

Eccleston, P. (2007) 1,300 tigers could be all that remain in the wild in India: Is the big cat doomed? *Daily Telegraph*, 30 August, 3.

Entwistle, A. and Dunstone, N. (eds) (2000) *Priorities for the Conservation of Mammalian Diversity: Has the Panda had its Day?* Cambridge University Press, Cambridge.

EIA (2004) *The Tiger Skin Trail*. Environmental Investigation Agency, London.

Fennell, D. (2001) A content analysis of ecotourism definitions. *Current Issues in Tourism*, 4, 403–421.

Foster, P. (2005) Tiger tourism hotels are taking the lion's share of the cash. *Daily Telegraph*, August 12, 15.

Goodwin, H. (2002) The case for responsible tourism. In T. Jenkins (ed.) *Ethical Tourism: Who Benefits?* Hodder and Stoughton, London.

Greenough, P. (2003) Bio-ironies of the fractured forest: India's Tiger Reserves. In C. Slater (ed.) *In Search of the Rainforest*. Duke University Press, Durham.

Guha, R. (1989) Radical American environmentalism and wilderness preservation: A Third World critique. *Environmental Ethics*, 11, 71–83.

Hannam, K. (2005) Tourism management issues in India's National Parks: An analysis of the Rajiv Gandhi (Nagarahole) National Park. *Current Issues in Tourism*, 8, 165–180.

Honey, M. (1999) *Ecotourism and Sustainable Development: Who Owns Paradise?* Island Press, Washington DC.

Ives, R. (1996) *Of Tigers and Men*. Doubleday, New York.

Karanth, K. (2006) *A View from the Machan*. Permanent Black, New Delhi.

Karanth, K. and Gopal, R. (2005) An ecology-based policy framework for Human–tiger coexistence in India. In R. Woodroffe, S. Thirgood and A. Rabinowitz (eds) *People and Wildlife: Conflict or Coexistence?* Cambridge University Press, Cambridge.

Levi, P. (1989) *Other People's Trades*. Michael Joseph, London.

Lewis, M. (2004) *Inventing Global Ecology: Tracking the Biodiversity Ideal in India 1947–1997*. Ohio University Press, Athens.

Luce, E. (2006) *In Spite of the Gods: The Strange Rise of Modern India*. Little Brown, London.

Matthews, J. (2000) A roar for life. *Discoverers*. Available at: www.discoveryinitiatives.com/articlestiger.asp [accessed: 9 January 2001].

Matthews, J. (2005) Tiger Tourism: Curse or Cure. *Discoverers*. Available at: www.globaltigerpatrol.co.uk [accessed 11 April 2007].

McCarthy, T. (2004) Nowhere to roam. *Time Magazine*, 23 August, 44–51.

Meacham, C. (1997) *How the Tiger Lost its Stripes: An Exploration into the Endangerment of a Species*. Harcourt Brace, New York.

Mills, S. (2004) *Tiger*. BBC Books, London.

Milton, K. (2002) *Loving Nature: Towards an Ecology of Nature*. Routledge, London.

Moulton, C. and Hulsey. E. (2002) *Kanha Tiger Reserve: Portrait of an Indian National Park* (2nd edition). Vakils, Feffer and Simons, Mumbai.

Mulder, M. and Coppolillo, P. (2005) *Conservation: Linking Ecology, Economics and Culture*. Princeton University Press, Princeton.

Mullan, B. and Marvin, G. (1999) *Zoo Culture* (2nd edition). University of Illinois, Urbana.

Padel, R. (2005) *Tigers in Red Weather*. Little Brown, London.

Project Tiger (2001) *State of the Tiger Report*. Project Tiger Directorate, New Delhi.

Project Tiger (2005) *Joining the Dots: Report of the Tiger Task Force*. Project Tiger Directorate, New Delhi.

Sarkar, S. (2001) Restoring wilderness or reclaiming forests? In D. Rothenberg and M. Ulvaeus (eds) *The World and the Wild.* Terra Nova, Newark.

Seidensticker, J. Christie, S. and Jackson, P. (eds) (1999) *Riding the Tiger: Tiger Conservation in Human Dominated Landscapes*. Cambridge University Press, Cambridge.

Singh, A. (1973) *Tiger! Tiger!* Jonathan Cape, London.

Sippy, S. and Kapoor, S. (2001) *The Ultimate Ranthambhore Guide*. Kapoor Publishing, Sawai Madhopur.

Terborgh, J. (1999) *Requiem for Nature*. Shearwater, Washington DC.

Thapar, V. (2006) *The Last Tiger: Struggling for Survival.* Oxford University Press, New Delhi.

Ward, G. and Ward, D. (2000) *Tiger Wallahs: Saving the Greatest of the Great Cats*. Oxford University Press, New Delhi.

Chapter 9
Managing Wildlife for People or People for Wildlife? A Case Study of Dingoes and Tourism on Fraser Island, Queensland, Australia

Georgette Leah Burns

Introduction

The title of this chapter immediately suggests that either people are managed for wildlife, or wildlife is managed for people. Suffice to say, neither is exclusively correct. In ecotourism settings, managers most commonly have to manage for both, in what is often a very complex balancing act. However, situations can arise where a choice of prioritizing management of one over the other has to be made, and the recent history of dingo and human interactions on Fraser Island provides a compelling example of this.

Dingoes and people have coexisted in Australia for thousands of years, and on Fraser Island for several hundred. More recently, this popular tourist destination was inscribed as a World Heritage Area in 1992, in recognition of its outstanding natural features, and consequently visitor numbers increased rapidly. Queensland Parks and Wildlife Service (QPWS) have managed the island, the wildlife, the tourists and their interactions since taking over from the Forestry Department in 1991. A common complaint by the government department managers is that their time is increasingly spent overseeing tourism issues. This detracts from their ability to address other important concerns on the island, such as weed control and wildfire prevention.

This chapter examines the impact of tourism on the dingoes of Fraser Island, and the aspirations of tourists in this nature-based tourism setting. As the uneasy relationship between wildlife and people on the island is explored, the crucial role of managers in this setting is highlighted. The case study investigates how management choices are made and what influences these decisions, before offering lessons for understanding factors affecting sustainable wildlife tourism in ecotourism settings.

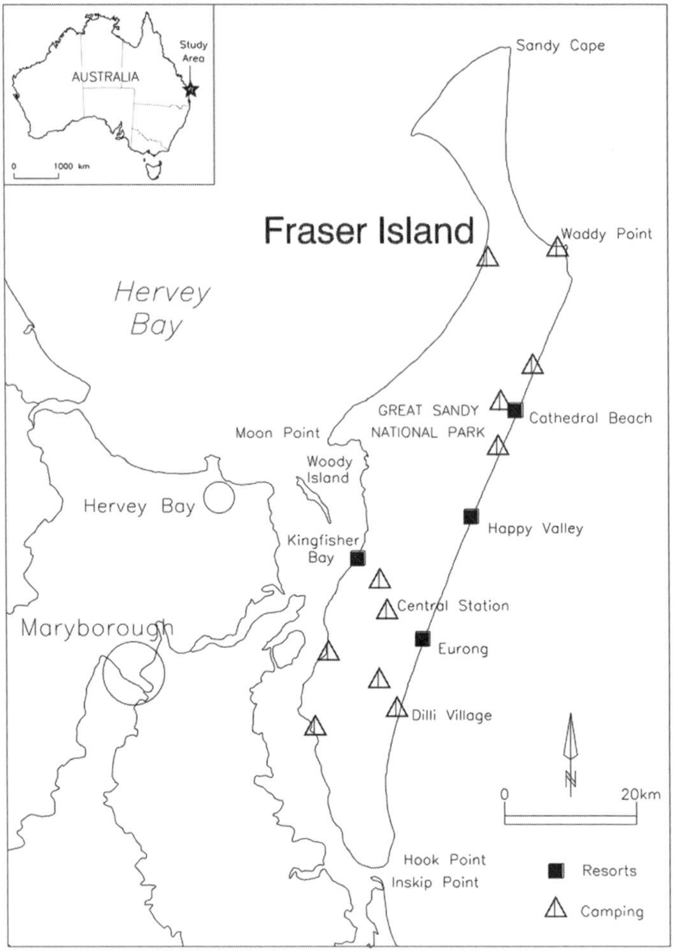

Figure 9.1 Fraser Island – Great Sandy National Park

Research Methods and Field Site

Methods

The voices heard in this chapter come from interviews undertaken on Fraser Island during five periods of fieldwork between 2001 and 2004. Literature is drawn from policy documents and newspaper articles relevant to the case study, as well as wider academic sources on wildlife tourism. This data is supplemented with observations of human and dingo behaviour on Fraser Island.

Guiding the research that informs this chapter are questions that ask why management of the human-wildlife interface happens in the way it does, and what values underpin the constructions of wildlife that, in turn, enable and legitimize the management. The case study of dingoes and tourism on Fraser Island provides a rich, contemporary setting within which to search for answers to these questions.

Fraser Island

Lying off the Queensland coast, approximately 400 km north of Brisbane, Fraser Island is the world's largest sand island. It is 123 km long, up to 25 km wide in some places and covers an area of 165,289 ha (Thompson et al. 2003). It has a long history of human resource exploitation through sand mining and logging. These activities were dominant in the 1960s and 1970s, but were actively opposed by community groups, such as the Fraser Island Defenders Organisation (FIDO) (Sinclair and Corris 1994), who were instrumental in obtaining World Heritage status for the island.

The Queensland Parks and Wildlife Service (QPWS) took over management of Fraser Island from the Queensland Forestry Department in 1991. It was World Heritage listed in 1992, in recognition of its unique natural features, and with this came a mandate to conserve the island's natural and cultural resources. The majority of the island (98 per cent) is a National Park, with small areas of freehold land in townships being the exception (Thompson et al. 2003). The Great Sandy Region Management Plan (1994–2010) guides the management of both Fraser Island and its surrounds (see Figure 9.1).

The Dingoes and other Wildlife of Fraser Island

The opportunity to view a wide variety of species is an important influence on visitors choosing Australia as a tourism destination (Rodger and Moore 2004), and is similarly important for visitors to Fraser Island where the size and ecological richness of the island supports a diverse range of wildlife. On the island there are over 350 different bird species ranging from birds of prey such as sea eagles (Shephard et al. 2005) and osprey, to common birds such as pelicans, honeyeaters and cockatoos, and vulnerable bird species such as the ground parrot (Chan and Mudie 2004). Hobson and Thrash (2005) have also recorded a rich diversity in the following reptile, amphibian and fish species: snakes (35), skinks (25), frogs (15), turtles (9), lizards (6), goannas (2) and freshwater fish (24). In addition to dingoes, other mammals on the island include swamp wallabies, echidnas, possums, sugar gliders and flying foxes. Brumbies (wild horses) were brought to the island for the logging industry in 1879, although the last of these were recently removed by QPWS (Thompson et al. 2003).

Dingoes were most likely introduced to Australia by Indonesian traders four to five thousand years ago, and came to Fraser Island with indigenous Australians

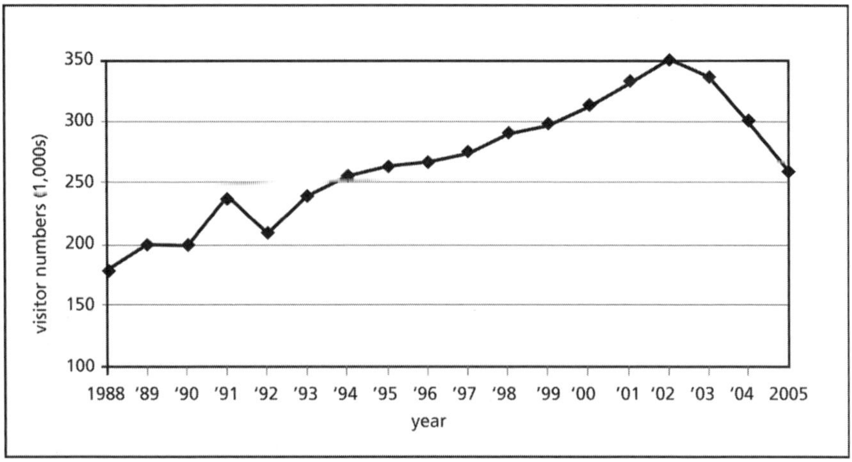

Figure 9.2 Visitor numbers to Fraser Island 1988–2005

Source: Compiled from Queensland Government Department of the Environment (1998) and annual QPWS figures.

at some point in time since then (Corbett 1995, Rogers and Kaplan 2003).[1] The dingo population on Fraser Island is thought to range seasonally between 100 and 200 animals (Corbett 1998) where, due to their isolation and limited interbreeding with other wild dogs, they are considered amongst the most pure in Australia (EPA 2001a, Thompson et al. 2003).

The People of Fraser Island

Fraser Island is alternatively home, work and playground to the many people who comprise its stakeholders. As many as 350,000 tourists visit the island annually (see Figure 9.2) and approximately 200 residents, living on freehold land, call the island 'home'. QPWS staff work and live there, as do resort owners and staff, and other tour operators. While each group has its own issues concerning wildlife and management (Burns and Howard 2003), the people focused on in this chapter are the tourists and the managers.

Many different types of tourists visit Fraser Island. The island is close enough to the mainland to be a destination for day-trippers from nearby mainland towns, such as Maryborough and Hervey Bay, and it is accessible by boat or aeroplane. It is a popular destination for both national and international backpackers who can hire four wheel drive vehicles or join tour groups on the mainland. Both budget and high end of the market accommodation is available on the island; in addition, there

1 Thompson et al. (2003) suggest dingoes have been present on the island for at least several hundred years.

are designated camping areas with visitor facilities, houses that can be privately rented, and several resorts.

World Heritage listing in 1992, and promotion of Fraser Island as pristine and unique, triggered a dramatic escalation in tourist and visitor numbers and a corresponding increase in management problems. Annual numbers tripled from 50,000 to over 150,000 in the late 1980s (Queensland Government Department of the Environment 1998), and peaked at 353,000 in 2002 (see Figure 9.2). Being a large island (see Figure 9.1), an analysis of carrying capacity may indicate that Fraser Island could continue to support greater numbers of tourists. The issue in this chapter, however, is with what these numbers have meant in terms of impacts on dingoes and shifting management responsibilities, as evidenced in policies and practices.

Tourism Impacts and Management

Tourism Impacts on Dingoes

As tourism increased on Fraser Island interactions with the dingoes also increased, and changed in nature. Little is recorded about dingo interactions with indigenous Australians on the island. We know that the relationship between dingoes and Aborigines on the mainland took many diverse forms (Meggitt 1965), with dingoes being viewed as variously wild and domesticated (the latter as companions, protectors and hunting dogs). Residents and workers present on Fraser Island during the 1960s and 1970s, when extractive resource use far outweighed tourism, recall accounts of befriending dingoes and dingoes stealing fish bait and garbage (Burns 2006) in what were considered typically benign interactions (Figure 9.3). Changes, then, seem to have occurred in parallel with increasing tourist numbers. QPWS recorded approximately 400 official reports of interactions categorized as 'negative' between the late 1980s and 2001 (EPA 2001a, Thompson et al. 2003), which ranged from dingoes snarling at tourists to the death of a 9-year-old boy in April 2001.[2]

The immediate and widely publicized response to this fatality was a cull of 31 dingoes, including the two involved in the fatal attack and 29 others that were known to frequent high visitor use areas (Thompson et al. 2003). While a large scale cull like this has not occurred since May 2001, dingoes identified as 'problem' animals continue to be culled on the island at the rate of approximately one per month (Ironside 2007). Although QPWS maintains there is little potential for long-term adverse impacts on the entire population (Thompson et al. 2003), culling has an extreme impact on individual animals. The need to cull is justified

2 Newspaper articles (Herald Sun 2007, Giles 2007) now regularly report a figure of '2000 incidents of menacing behaviour' by dingoes toward humans on Fraser Island since April 2001.

Figure 9.3 Dingo attracted to fishing gear on Fraser Island

by QPWS's conviction that dingo attacks are a consequence of habituation, which itself is a direct result of humans feeding the animals (EPA 2001a, Thompson et al. 2003).

Fraser Island Policy Documents

Prior to the death in 2001, QPWS had been managing Fraser Island primarily for the purpose of conservation of its flora and fauna. However, QPWS has now been forced to prioritize wildlife management for tourism recreation, and it is not a practice that sits easily with the department or the on-ground managers. The dilemma of management has become increasingly apparent in works such as Thompson et al. (2003: 37), all employees in QPWS at the time of the fatality, who ask in their abstract 'Do we manage the animals or the people?'

This dilemma is immediately apparent in the Fraser Island Dingo Management Strategy (FIDMS) that notes: 'On Queensland national parks the dingo is protected as a native species and the QPWS has a legal responsibility to conserve these populations even though the dingo is declared a pest in this State' (EPA 2001a: 4). This imperative comes from the rather ambiguous legal situation across Australia that deems dingoes as protected within national parks while simultaneously declared a pest across most of the rest of the country (Hytten and Burns 2007a),[3, 4] The necessity of their protection on Fraser Island implies that successfully managing to conserve this species should be a priority for QPWS and, indeed, the first of five stated overall objectives of the FIDMS is to 'ensure the conservation of a sustainable wild dingo population on Fraser Island' (EPA 2001a: 4).

The FIDMS also recognizes, however, that 'the common law duty of care requires the QPWS to address the safety of people entering and using the lands, water and facilities in the Fraser Island Recreational Area' (EPA 2001a: 6). Thus, the safety of people is also a priority and, in a context where the wildlife has attacked and killed people, this clause cannot be ignored. In addition, the FIDMS also recognizes that '[t]he QPWS has obligations to ensure that management is ethical and humane' (EPA 2001a: 6). Put into practice, this policy means that dingoes must not kill people, but people (the QPWS managers of the island) can kill dingoes to ensure human safety, as long as the dingoes are killed in a way that is deemed ethical and humane.

The second significant policy document governing management of the island, its wildlife and the people who visit or reside on it, is the Great Sandy Management Plan (GSMP) (Queensland Government 2005). A review of changes in the GSMP over time reveals necessary alterations in the way the plan deals with people. The GSMP was first released in 1994 and opens with a statement from the then Minister for Environment and Heritage, Molly Robson, who was very optimistic about the opportunity created by World Heritage listing on Fraser Island to 'preserve [the] natural beauty' of the region (Queensland Government 2005: 3). The Minister's message held a grand vision for implementation of the plan; that the area it covered could be everything for everybody:

> … a place where nature's evolutionary processes can continue unimpeded, a bequest to all humanity, where forests can regenerate and flourish, where the freshwater lakes contain pure unpolluted rainwater, where the sand dunes built up over the ages continue to tell their story of geological wonder, where Indigenous heritage and aspirations can find true expression, where residents can earn their living, and where tourists … can enjoy its splendour and tranquillity and return home without having marred their priceless heritage (Robson 1994, in Queensland Government 2005: 3).

When the plan was revised in 2005, tourists marring heritage was of much less concern than tourists going home unharmed by wildlife. While tourism was

---

3 Dickman and Lunney (2001: 96) label this 'the kill-or-conserve dilemma.'

4 The legalised protection of dingoes is only relatively recent, as up until the 1970s they were almost universally treated as vermin (Hytten and Burns 2007b). Call for a widening of the protection of this species was made by the CSIRO in the 1980s (The Age 2007) and has since been re-voiced by wildlife researchers (Cooper 2007) and the Humane Society International (ABC 2007).

clearly an important and growing concern for the region in 1994, its centrality was more evidenced in the revised plan. Here, the opening message from the Minister for Environment, Desley Boyle, talks positively about progress that has been made since the initial plan, notes an increase in 'tourism management policies for protected areas of the region', and explicitly mentions visitor safety and dingoes: 'Safety is a key concern, and a campaign has raised awareness, especially in relation to averting further problems with dingo attacks' (Boyle in Queensland Government 2005: 4).

Tourist Aspirations

On tour company web pages Fraser Island is marketed as a peaceful paradise with an abundance of unique flora and fauna (for example, Fraser Island Company 2007, Kingfisher Bay Resort 2007). Dingoes are a main icon of the island and feature heavily on most web pages and other marketing avenues, such as brochures and postcards. The FIDMS (EPA 2001a: 4) notes that the close and obvious presence of dingoes 'is a tourist attraction and marketing drawcard'.

From interviews with tourists on the island it is clear that a key aspiration for most of them is to experience the natural beauty of the location, and this is done in a variety of ways, without being hassled by external forces. Like tourists in most destinations, holidays are valued for the freedom they provide from the constraints of everyday life. Tourists want to relax and get away from the hectic pace of their normal day to day existence (Cohen 1972). In choosing a national park as their destination, many are keen to immerse themselves in that experience and be a part of that natural setting. On Fraser Island the tourism is very much nature-based, although it does not always adhere to the principles of conservation promoted in ecotourism guidelines (for example, The International Ecotourism Society 2007).

Answering questions about why they were visiting the island, tourists responded in the following ways:

- Just wildlife, camping and beaches – a bit of sun.
- Having a holiday, [we have] always wanted to come here.
- Just to see whatever happens, and [experience the] wildlife.
- [Fraser Island is] very special – it's very religious, sacred – you can sleep well at night and you can walk well with a natural stride – just no panic in your daily activities – it's a relaxed atmosphere.
- That's the beauty of the place. We can come here and do a little bit of fishing, photography or just relax.
- … just the dingoes, and rainforest.
- Wildlife, camping, enjoy[ing] the environment.
- The basic wildlife that is available – that's probably the main thing – and that it gives you a taste of being away from civilisation.

In addition, people come to see what they have been informed through marketing campaigns to expect:

> I wanted to come because of all the nice things, the sands and all the things you hear about Fraser Island … and because it's got World Heritage listing and it's pure and nature and stuff.

How do these, very normal, tourist expectations of their stay on the island compare with what managers want and expect, and with wildlife needs?

Manager Expectations

There is a very evident frustration by Fraser Island managers, particularly those 'on the ground', that has increased since I first began collecting data on the island in 2001. It became apparent at a conference hosted by the Fraser Island Defenders Organisation (FIDO) on 'The Integrity of Fraser Island' in 2006, where a common complaint from managers was that increasingly their time and resources are consumed by recreation management.[5] This impedes their ability to deal with other pressing management issues, such as wild fire prevention and weed control.

The change in management focus is a consequence of the large number of tourists visiting the island and the fatality, for which Thompson et al. (2003: 38) note QPWS were unprepared and about which they were 'surprisingly naïve'. Such naïvety may not rest solely with QPWS, but is perhaps also characteristic of wildlife management in other parts of Australia. It is important to note that human deaths, as a result of interactions with wildlife in national parks, have been uncommon in Australian history. The country has no large terrestrial carnivores, unlike national parks in Africa or North America.

QPWS rangers on Fraser Island described their work prior to the fatality in the following ways:

> It's basically just the general conservation issues … preservation to the greatest possible extent.

> Looking after the park as far as estate maintenance and development, infrastructure, natural resource management, be it fires, weed control, stuff like that.

They saw their role as one of managing the wildlife in the context of the entire ecosystem, for the overall conservation of the location as a whole system. In this view, wildlife is constructed as part of other natural features, such as vegetation.

5 Worboys et al. (2005) note that managing protected areas to minimise risks to visitors while maintaining a high quality recreational product requires considerable resources.

However, increasingly there was recognition that managing people was becoming dominant:

> Recreational management tends to overtake just about everything else … that's where the issues are and that seems to be overriding in this location. We do a million different jobs; whatever comes, we do it basically. It's all the infrastructure that goes with it for a start; toilets, boardwalks, recreation areas, camping grounds, development and maintenance of those. Visitor management is the other big side of it, just getting out and talking to people, law enforcement stuff, education, interpretation programmes, that's what takes up most of the time. I guess the thing about this place is it's [a] National Park, it's also [a] World Heritage area, it's also [a] recreation area, so you're trying to manage under three distinct regimes. The money comes from the recreation side of it, so that's where the money goes back to – that seems to be the overriding push … on this particular location anyway.

Managing Interactions

Following a risk assessment (EPA 2001b) and final version of the FIDMS (EPA 2001a), QPWS introduced a multi-strategy approach aimed towards minimizing human–dingo interactions. The purpose was to encourage both people and dingoes to behave in the manner QPWS considered appropriate; that is, not to interact. To discourage humans from seeking interactions with dingoes, QPWS increased their educational campaign and interpretation material, and had it evaluated by an external consultancy (QPWS 2003a, 2003b) who concluded it was equivalent to best practice in North American parks.

The strategy also involved an increase in, and greater enforcement of, fines (ranging from $A225 on-the-spot to a maximum of $A3000 court-imposed for feeding dingoes). Fencing of areas categorized as 'high risk', such as camping grounds and resorts, was also included and is about to increase. The Premier of Queensland, Anna Bligh, announced in September 2007 that the state government would spend $A750,000 erecting more fences on the island (Herald Sun 2007).

The fencing is also designed to discourage dingoes from interacting with people. Other current strategies for managing dingoes include culling, as discussed earlier, and hazing. Hazing is a type of aversion therapy, which involves shooting dingoes with ratshot from guns or clay pellets from slingshots (Giles 2007). In addition, an audiosonic deterrent was trialled in two locations on the island in 2003, but it did not succeed in excluding dingoes from the trial areas (Edgar 2004).

An obvious way to avoid further interactions between people and dingoes is to remove one or the other entirely from the island. Given the anthropocentric values underpinning wildlife management, the fact that some 200 people reside on the island, and the considerable financial input from tourists, total removal of people has not been discussed in policy documents. An early proposal in the FIDMS (EPA 2001a: 11), however, promised to investigate 'the possibility of limiting visitor numbers to the island or at specific locations on the island'. As a consequence, some camping areas have been closed, but there has been no cap on numbers.

Total removal of dingoes has, similarly, not been enacted. As noted earlier, the FIDMS formally recognizes the dingo as a protected native species on the island and therefore the assumption is made that dingo conservation is warranted. While some wildlife researchers consider the species an introduced pest (for example, Monaghan, in Green 2005, Van Dyck 2007), there are also some who increasingly support its conservation (for example, Dickman, in Cooper 2007), and in May 2007 the Department of Sustainability and Environment issued a 'preliminary recommendation' of the dingo as an endangered species (The Age 2007). If this trend towards positively constructing the dingo as a valuable part of contemporary Australian wildlife continues, it would be reasonable to expect increased public and scientific pressure against the regular culling of Fraser Island dingoes. This pressure is not without precedent, as the large scale cull triggered by the death in 2001 became a hotly debated issue (Burns and Howard 2003). Such pressure would undoubtedly cause more frustration for Fraser Island managers who, in general, see their roles as conservationists and not cullers.[6]

As discussed earlier, the legislation in Australia surrounding dingoes is confusing, and this does not assist management clarity. Dingoes are variously seen as a pest versus a protected species, a native versus an introduced species and a pure-bred versus a hybrid species. From this, it can be argued that the approach to dingoes is related to how they are socially constructed. If they are recognized as protected, native and pure, then there is less incentive to cull them. If, however, they are considered as introduced, pest and hybrid, then it is more acceptable to cull them to remove the chance of contact with people (Hytten and Burns 2007a, 2007b).

Re-Constructing People and Nature: Re-Constructing Management

Although there has been a decrease in tourist numbers on Fraser Island since 2002 (Figure 9.2) it remains a popular tourist destination and this is unlikely to change. There is a need, therefore, for a management shift, both conceptual and practical, particularly in relation to people. Management practice has been guided by ideological thinking reminiscent of colonialization periods when humans and nature were seen as separate and humans able to control nature (Suchet 1999,

6 This view is not universally shared. Howard (2007), for example, argues that wildlife management in Australia has grown out of pest management and tensions arise when species require conservation, particularly if that species has been labelled a 'pest' from the onset of European colonization. However, these views are compatible if we accept that the tensions exist both within the wildlife management profession and between wildlife managers and wildlife 'consumers' (in this case Fraser Island visitors) based on the different ways that these groups construct wildlife.

Adams 2003). Discourse in the FIDMS (EPA 2001a) blames dingo attacks on habituation that results from people feeding dingoes,[7] and construction of humans as interfering with nature in this way suggests we are something other than nature (Aitken 2004: 53). This is reflected in current strategies, such as fencing, that reinforce the separation of people and nature.

The shift could come in recognizing people as an integral part of, rather than separate from, the system that QPWS manages. Current tourism and management literature advocate a holistic systems approach in theory (Burns 2004, Worboys et al. 2005), which takes into account all variables and is, therefore, inclusive of both people and wildlife.

While still focusing on the separateness of people and wildlife, the educational focus in the FIDMS encourages managers to become interpreters of the wildlife experience for tourists, and not just mediators or control agents. This has taken the form of increased literature for tourists to read, including brochures and signage on the island, and contact with rangers at camp sites and as visitors disembark from ferries. This increasing recognition of the need to manage people is important because not managing people well enough frequently leads to a need to act on anthropocentric values and punish wildlife for being 'evil' (Howard 2007).

There are, however, problems with the QPWS 'be Dingo-Smart' educational message. Firstly, it is difficult to ensure the message is received and understood and, secondly, the message comes from a particular stance that is informed by values that are not shared by all (Hytten and Burns 2007b). There is also the danger that too much interpretation could turn visiting Fraser Island into a zoo-like experience, which is not expected or wanted by tourists in this setting. Tourists want peace and relaxation, and natural beauty that comes with their choice of ecotourism setting in locations such as Fraser Island. Fencing and noticeboards with warnings and messages are the antithesis of that.

One of the key characteristics that sets ecotourists apart from other tourists is their desire to experience, and be a part of, natural settings. Thus, ecotourism, more than any other form of tourism, offers the opportunity to manage in a holistic framework that includes people as a part of nature. Tourists in natural settings should be encouraged to see themselves as part of the environment and made aware that their actions within it can have far-reaching consequences, both for themselves and for the wildlife. Similarly, in this setting wildlife should be permitted to behave in ways that are natural with awareness by people that this may have consequences, both positive and negative, should they interact. Constructing the ecotourism setting in this way transfers a sense of responsibility to tourists. Managers, rather than assuming control of both people and wildlife, are relieved of some of the increasing burden of recreational management.

---

7 This relies on an argument that habituation is not natural and interactions between people and dingoes are not natural, yet both have occurred for as long as there has been contact between people and dingoes.

Encouraging tourists to be aware of their responsibility, for themselves and to nature, may be a positive step forward in ecotourism settings. If tourists are aware that dingoes, wildlife and other forms of nature belong in the ecotourism setting, then they are more likely to be accepting of their behaviours. Taking this further, attacks from wildlife would be constructed as normal. Hunting and aggressive play is natural behaviour for dingoes (Corbett 1995) and, because they are wild animals, tourists should not expect them to behave in the same way as domesticated ones. If attacks are perceived as a normal, rather than abnormal, part of the reality of some nature-based settings, then tourists face the decision of whether to visit the destination or go elsewhere. The responsibility is theirs.

Rather than focusing on managing people for wildlife or wildlife for people in a construction that separates people from nature, to maintain the integrity of the wildlife tourism setting managers should promote awareness of people as part of the setting. This has the potential to remove some of the anthropocentric dominance that underlies management and views of nature. Constructed in this way, dingoes frequenting high-use human areas such as campgrounds and resorts are doing something natural, rather than crossing a known line between human and animal spaces. Consequently, this is not a problematic behaviour and does not warrant the animal being culled. Similarly, this construction recognizes that Fraser Island is not a zoo, and fences that separate people and nature are not natural. There is potential for this approach to be successful on Fraser Island and in other nature-based tourism settings where ecotourists are more likely to be accepting of natural features, and dangers, of an area and, therefore, easier to manage in this 'hands off' scenario.

Conclusion

For QPWS to achieve sustainable wildlife tourism on Fraser Island, in accordance with the guidelines for world heritage listing and principles for ecotourism, they need to effectively manage both the wildlife and the people who interact with that wildlife. Attempts to do this have focused on reducing all interactions regardless of whether or not they may be viewed as desirable or undesirable. Interpretive material aimed at educating tourists and other island visitors or residents about the negatives of wildlife interactions (where the dominant message is that the people might get hurt and the wildlife will get culled) has increased. Following a study commissioned to evaluate the current material and suggest ways it might be improved (Beckmann and Savage 2003, QPWS 2003a, 2003b), much larger fines were introduced for human misdemeanours (such as feeding, either directly or indirectly), and their enforcement publicized through local media (for example, Green 2005, Williams 2006). On the wildlife side, fences continue to be constructed to minimize interactions, and the 'problem' wildlife culled in accordance with the FIDMS. But clearly these interventions are not working. The dingoes are still a nuisance and are still being culled. The tourists are staying away, and the managers are frustrated by this ongoing, time-consuming issue.

Time and management training will hopefully see a shift in focus that ensures park managers expect to manage people as a crucial part of the natural system, and this may alleviate some of their frustration. The recordings of negative dingo interactions are increasing and, in line with the current policy of culling problem animals, this means the rate of dingo deaths is likely, if anything, to increase. Why tourist numbers are decreasing remains to be investigated. Part of finding a way to ensure the sustainability of wildlife tourism on Fraser Island lies in being more critical of management and closely examining, and acknowledging, for whom it is being done.

Fraser Island is an extreme case, as not all interactions between wildlife and people result in people being harmed and wildlife being culled. However, there are lessons that can be taken from this case that are applicable to others. In all wildlife tourism settings, the needs of both people and wildlife must be understood before they can be effectively managed. When the aspirations of tourists are clear and the impact on wildlife is obvious, there is a need to acknowledge who is being prioritized in management and why. The aims of the management need to be transparent, and the values and constructions that underpin management choices and decisions need to be recognized (Hytten and Burns 2007b).

'Walking the management tightrope between keeping wildlife wild and tourists safe' (Thompson et al. 2003: 46) requires complex balancing of management priorities. Strategies for managing dingoes and humans on Fraser Island attempt to manage people for wildlife and wildlife for people, but there is a clear dominance of one over the other. This is not uncommon. Where wildlife is perceived to threaten humans, managing for the protection of humans will always be prioritized. Legislation that then enacts the culling of wildlife for fear of people being harmed, demonstrates an obvious anthropocentric view of management that puts humans first.

Managers have a crucial role to play in sustainability. This has long been recognized, in parks and other protected areas, with regards to the conservation of natural features. Traditionally this occurred in a context in which those natural features were deemed to be separate from human ones. Separating humans from nature only exacerbates the problem, and certainly is not proving to be effective in terms of sustainable tourism on Fraser Island. Consequently, one way to manage more effectively might be to construct people as part of the overall natural system, and not separate from it. This is difficult to implement in practice, as most problems are easier to conceptualize as isolated variables and then manage as such. However, once constructions are recognized and people re-categorized, we may be able to see a way forward, not just on Fraser Island but in all wildlife tourism settings.

Is it possible to manage in the best interests of both people and wildlife, and keep both happy? Clearly for wildlife to be sustainable this is what managers must aim to achieve, and the success obviously depends on many factors. Crucial, however, is an understanding of what tourists want, coupled with what wildlife needs, and why certain management decisions are made. Locating people in nature and shifting some of the responsibility from managers to tourists may reduce the heavy burden of recreation for managers, *and* provide more satisfaction for tourists, greater conservation values for wildlife, and an enhanced wildlife tourism experience for all.

References

ABC (2007) Humane Society urges dingo protection. *ABC News* [Online]. Available at: www.abc.net.au/news/stories/2007/05/24/1932203.htm [accessed: 18 October 2007].

Adams, W.M. (2003) Nature and the colonial mind. In W. Adams and M. Mulligan (eds) *Decolonizing Nature: Strategies for Conservation in a Post-Colonial Era*. Earthscan, London, pp. 16–50.

Aitken, G. (2004) *A New Approach to Conservation: The Importance of the Individual Through Wildlife Rehabilitation*. Ashgate, Aldershot.

Beckmann, E.A. and Savage, G. (2003) *Evaluation of Dingo Education Strategy and Programs for Fraser Island*. Prepared for Queensland Parks and Wildlife Service. Environmetrics, North Sydney, New South Wales.

Burns, G.L. (2004) Anthropology and tourism: past contributions and future theoretical challenges. *Anthropological Forum*, 14, 5–22.

Burns, G.L. (2006) The fascination of fur and feathers: managing human–animal interactions in wildlife tourism settings. *Australian Zoologist*, 33, 446–457.

Burns, G.L. and Howard, P. (2003) When wildlife tourism goes wrong: a case study of stakeholder and management issues regarding dingoes on Fraser Island, Australia. *Tourism Management*, 24, 699–712.

Chan, K. and Mudie, D. (2004) Variation in vocalisations of the ground parrot at its northern range. *Australian Journal of Zoology*, 52, 147–158.

Cohen, E. (1972) Towards a sociology of international tourism. *Social Research*, 39, 164–182.

Cooper, D. (2007) Dingoes touted as wildlife's saviour. *ABC Science* [Online]. Available at: www.abc.net.au/news/stories/2007/07/12/1976539.htm [accessed: 18 October 2007].

Corbett, L. (1995) *The Dingo in Australia and Asia*. University of New South Wales Press, Sydney.

Corbett, L. (1998) *Management of Dingoes on Fraser Island*. ERA Environmental Services Pty Ltd., Report for Department of Environment and Heritage.

Dickman, C.R. and Lunney, D. (2001) Last howl of the dingo: the legislative, ecological and practical issues arising from the kill-or-conserve dilemma. In C.R. Dickman and D. Lunney (eds) *A Symposium on the Dingo*. Royal Zoological Society of New South Wales, Mosman, pp. 96–107.

Edgar, J. (2004) *The Effectiveness of Auditory Stimuli for Canid Deterrence.* Unpublished honours thesis. Faculty of Environmental Science, Griffith University, Brisbane.

EPA (Environmental Protection Agency) (2001a) *Fraser Island Dingo Management Strategy.* Conservation Management Report. Queensland Parks and Wildlife Service, Brisbane.

EPA (Environmental Protection Agency) (2001b) *Risk Assessment – Risks to Humans Posed by the Dingo Population on Fraser Island.* Conservation Management Report. Queensland Parks and Wildlife Service, Brisbane.

Fraser Island Company (2007) *Welcome to Fraser Island.* Available at: http://www.fraserislandco.com.au/ [accessed: 18 October 2007].

Giles, D. (2007) New sting in dingo fight. *The Sunday Mail*, 24 June 2007.

Green, G. (2005) More dingo victims 'inevitable'. *The Courier Mail*, 17 June 2005.

Herald Sun (2007) Fraser Island gets new dingo fences. *Herald Sun*, 23 September 2007.

Hobson, R. and Thrash, I. (2005) *Reptiles, Frogs and Freshwater Fish: Fraser Island World Heritage Area.* Fact Sheet. EPA, Brisbane.

Howard, P. (2007) *The Beast Within: An Exploration of Australian Constructions of Wildlife.* PhD thesis. Griffith University, Brisbane.

Hytten, K. and Burns, G.L. (2007a) *Dingo Dualisms: Constructions of Nature and Culture in the Australian Context.* Paper given at the Animals and Society Conference, Tasmania, July 2007.

Hytten, K. and Burns, G.L. (2007b) Deconstructing dingo management on Fraser Island, Queensland: The significance of social constructionism for effective wildlife management. *Australasian Journal of Environmental Management*, 14, 40–49.

Ironside, R. (2007) Rangers kill dingo. *The Courier Mail*, 20 April 2007.

Kingfisher Bay Resort (2007) Kingfisher Bay Resort Official Website. Available at: http://www.kingfisherbay.com/kingfisher/home/home_home.cfm [accessed: 10 October 2007].

Meggitt, M.J. (1965) The association between Australian aborigines and dingoes. In A. Leeds and A. Vayda (eds) *Man, Culture and Animals.* American Association for the Advancement of Science, Washington, DC, pp. 7–26.

Queensland Government (2005) *Great Sandy Region Management Plan 1994–2010.* Revised version September 2005. EPA, Brisbane.

Queensland Government Department of the Environment (1998) *Managing Visitor and Commercial Options for Ecological Sustainability: Final Report of the Review of Tourism Activities in the Great Sandy Region.* Department of the Environment, Brisbane.

Queensland Parks and Wildlife Service (QPWS) (2003a) Who got the message? Measuring the effectiveness of the communication strategy for dingo awareness on Fraser Island. *Queensland Parks and Wildlife Service Staff Newsletter*, 39, 8–9.

Queensland Parks and Wildlife Service (QPWS) (2003b) Our dingo education best in world *Queensland Parks and Wildlife Service Staff Newsletter*, 39, 1.

Rodger, K. and Moore, S. (2004) Bringing science to wildlife tourism: the influence of managers' and scientists' perceptions. *Journal of Ecotourism*, 3, 1–19.

Rogers, L.J. and Kaplan, G. (2003) *Spirit of the Wild Dog: The World of Wolves, Jackals and Dingoes*. Allen and Unwin, Crows Nest.

Shepard, J., Hughes, J., Catterall, C. and Olsen, P. (2005) Conservation status of the White-Bellied Sea-Eagle *Haliaeetus leucogaster* in Australia determined using mtDNA control region sequence data. *Conservation Genetics*, 6, 413–429.

Sinclair, J. and Corris, P. (1994) *Fighting for Fraser*. Kerr Publishing, Alexandria.

Suchet, S. (1999) *Situated Engagement: A Critique of Wildlife Management and Post-Colonial Discourse*. Unpublished PhD thesis. Macquarie University, Sydney.

The Age (2007) Pest of the past, dingo's star in the ascendancy. *The Age*, 22 July 2007.

The International Ecotourism Society (2007) *Definitions and Principles*. Available at: http://www.ecotourism.org [accessed: 18 October 2007].

Thompson, J., Shirreffs, L. and McPhail, I. (2003) Dingoes on Fraser Island: tourism dream or management nightmare. *Human Dimensions of Wildlife*, 8, 37–47.

Van Dyck, S. (2007) Dingoes are not native dogs. *The Australian*, 25 April 2007.

Williams, B. (2006) Dingo nips toddler. *The Courier Mail*, 23 November 2006.

Worboys, G.L., Lockwood, M. and De Lacy, T. (2005) *Protected Area Management: Principles and Practices* (2nd edition). Oxford University Press, South Melbourne.

Chapter 10

Kohlberg: A Theoretical Framework for Informing Seal-Viewing Behaviour at the Kaikoura Peninsula, New Zealand

Davina Stanford

Introduction

Tourists are increasingly searching for more experiential types of tourism, and ecotourism is one such example. However, the term ecotourism can be misleading with the prefix 'eco-' suggesting a type of tourism that is beneficial to the environment; poorly managed ecotourism, like any other type of tourism, can cause negative impacts (Wheeler 1994, Boyd and Butler 1996, Mann 2000). What is more, ecotourism draws tourists into fragile, remote and marginal areas that may be more vulnerable to the impacts of tourism. Evidently, such visits need to be managed and controlled if the visitors are not to disturb and damage the places and wildlife that they are visiting. Cater and Goodall (1997: 88) write that tourists should understand a destination's 'sense of place' in order to respect its environment and culture. The key to this understanding is thought by many to be education and the dissemination of knowledge (Krippendorf 1984, Gunn 1988, Eber 1992, Prosser 1992, Forsyth 1996, France 1997, Reisinger 1997, Luzar et al. 1998, Boniface 1999, Broadhurst 2001).

Aimed at informing the tourist about the sites they are visiting, enhancing their experience, and guiding the visitor in terms of appropriate behaviour, is a proliferation of signage, codes of conduct, interpretation and so on at tourism and ecotourism sites. Despite its importance in managing visitor behaviour, much of this information is ad hoc and there is little empirically tested theoretical foundation for the content of the messages that are distributed to visitors. This chapter outlines some theoretical approaches which could be used in the provision of information that is intended to guide or inform appropriate behaviour. In particular, various ethical theories are considered and Kohlberg's Stages of Moral Development are explored as a possible conceptual framework to inform the development of information for visitors that is intended to guide and encourage appropriate behaviour. The theory is then applied empirically, based on a case study of the fur seal colony at Kaikoura Peninsula, on the South Island of New Zealand. Conclusions are made regarding the use of Kohlberg's Stages of Moral

Development, which emphasize the importance of positive, credible and reasoned messages.

Information and Communication

Prerequisites for Effective Information and Communication

Before looking at the empirical work presented here, we should consider the existing literature in the field of visitor information. There are several factors that have been identified as contributing to the effectiveness of communication. While this chapter is not intended to cover the literature extensively on this subject, before looking more closely at the means and content of communication it would be useful to revisit some of the prerequisites for successful communication of messages intended to guide appropriate behaviour:

- Of primary importance is the behaviour that is being targeted, as information will only be effective in guiding certain behaviours. Behaviour such as careless action (for example littering), unskilled action (for example selecting an improper campsite) or uninformed action (for example using dead snags for firewood) may be amenable to communication (Manning 1999). Communication may not be so effective, however, for behaviours such as illegal activities (for example graffiti) or unavoidable activities (for example disposal of human waste) (Manning 1999).
- Effective communication is also dependent upon multiple channels or media, and strategies to encourage desired recreational behaviour patterns must include a wide range of management interventions in addition to communications initiatives (Mason and Mowforth 1995, Carter 2001a).
- Information needs to be easily understood, should be well disseminated and widely promoted (Mason and Mowforth 1995).
- Novelty, interaction, activity and personal interest are found to be effective for interpretative messages, while repetition has been found to be related to decreased visitor attention (Moscardo 1996).
- Finally, the values of the recipients themselves will affect the success of communication.

The Means of Communication

In his seminal work, *Interpreting Our Heritage*, Freeman Tilden (1977: 3) describes interpretation as '… revealing, to such visitors as desire the service, something of beauty and wonder, the inspiration and spiritual meaning that lie behind what the visitor can with his senses perceive'. More recently, and somewhat less romantically, interpretation has been seen as a useful tool in visitor management that can modify and influence the behaviour of visitors (Manfredo 1992, Roggenbuck 1992, Hall and McArthur 1993, Moscardo 1998, Carter 2001a, 2001b). Interpretation can relieve crowding and congestion, alter behaviour directly by giving information or indirectly by fostering appreciation, and can create public support for conservation (Moscardo 1996).

Interpretation is more than simply providing written information and is, in essence, about stimulating visitors' imaginations or emotions so that they engage in a positive way with the places that they visit through a variety of media and communication channels (Carter 2001b). Although interpretation is upheld as a means for the effective management of visitors, Carter (2001a, 2001b) documents examples where interpretation has failed to produce the desired outcomes in terms of influencing visitor behaviour.

Codes of conduct attempt to promote a more responsible form of tourism (Mason and Mowforth 1996). Codes are produced by governments, communities, NGOs, religious and environmental groups and by industry and can be aimed at industry, government, hosts and tourists alike (Mason and Mowforth 1996, Malloy and Fennell 1998). Responsibility is manifest in two particular areas with codes targeting the environment and the culture of the host region (Mason and Mowforth 1996). One problem with codes of conduct in general is that they can appear admonitory or patronising, detracting from the relaxation of the holiday period (Mason and Mowforth 1995). Forsyth (1996: 14) quotes one tour operator who comments that people do not want to go on holiday with their school teacher. Butcher (2003: 141) also criticizes codes of conduct for spoiling the fun of holidays: 'Attempts to formalize codes of conduct, and the constant appeals for deference to the interests of the host's 'environment' and 'culture' only contribute to a spirit of caution rather than one of adventure and discovery'. Furthermore, behavioural control through sanctions may have an adverse effect on the quality of visitors' experience (Carter 2001a) and recipients of messages are believed to be more receptive to positive statements than to negative ones (Malloy and Fennell 1998). Clearly, the way in which codes of conduct are presented is crucial to the way in which these messages are received, and the content of codes of conduct is an area that requires further research (Mason and Mowforth 1996, Malloy and Fennell 1998).

Theoretical Frameworks for the Communication of Messages to Guide Appropriate Visitor Behaviour

Ethics: Consequentialism and Deontology

There are many branches of ethical reasoning, but two in particular are dominant schools of thought: consequentialism and deontology. Although not the only moral theories that philosophers have put forward, they are relevant to the context of guiding appropriate tourist behaviour because the principles they offer are addressed clearly to individuals and are intended as a basis for decision-making, which helps to identify the right thing to do in any particular scenario (Thomson 1999). Under the banner of consequentialism are further sub-divisions of philosophical thought: hedonism and utilitarianism. Simply put, hedonism is concerned with maximizing pleasure and minimizing pain for the individual whereas for utilitarianism,[1] the primary determinant of ethical conduct is the greatest good for the greatest number that results from one's actions (Brody 1983). Under the utilitarian branch of ethics are two further distinctions: act and rule utilitarianism. Act utilitarianism is focused solely upon the ends without considering the means, whilst rule utilitarianism argues that the greatest good for the greatest number must be achieved by following prescribed rules and acceptable means (Thomson 1999). Rule utilitarianism is thought to be preferable for organizing principles and rules of desired behaviour and for communicating messages targeting behaviour. This is because a rationale for an action is indicated (Malloy and Fennell 1998).

Deontology,[2] by contrast, proposes that certain actions are right or wrong and this approach describes ethical conduct as that which is in accordance with rules or the right means regardless of the consequences (Brody 1983). Deontology is often contrasted with consequentialism. Deontology would teach that the act of taking an innocent human life was wrong, no matter what the reason, while utilitarianism would recognize that innocent human life should not arbitrarily be taken, although it could be acceptable if taking that life was necessary to save the lives of many other innocent people. In its somewhat more prosaic application with regards to guiding tourism behaviour, deontology would not provide a rationale or justification for a certain action other than something being an obligation that one ought do as one's duty (Malloy and Fennell 1998). For example, a sign saying 'stay on the path', may be described as deontological, as it does not explain why the action is required whereas a rule utilitarian sign may state 'stay on the path; trampling can damage fragile plants', thus providing a rationale for the action.

Deontological ethical reasoning is evident in the codes of conduct analysed by Malloy and Fennell (1998). Their analysis of 40 codes comprising 414 statements shows that codes of conduct are generally deontological in nature (the recipient is told to do something because they should, rather than providing the rationale for this action). However, research in the area of management ethics has emphasized the importance of rule utilitarianism approaches to education, where an individual can learn through understanding the consequences of their actions (Malloy and Fennell 1998). The influence of deontological versus consequentialist ethics on the behaviour of the tourist is certainly worth applying beyond codes of conduct: for site-specific interpretation, for information about responsible tourism in general and for understanding motivations for responsible action.

---

1 Originated by Jeremy Bentham in the late eighteenth century and advocated by John Stuart Mill in the nineteenth century.

2 Developed by Immanuel Kant in the late eighteenth century in opposition to the principles of utilitarianism.

Table 10.1 Kohlberg's Stages of Moral Development

	Stage of moral development	Reasons for doing right
	Pre-conventional morality	
1.	Fear of punishment	Avoidance of punishment and the superior power of authorities.
2.	Maximizing pleasure/minimizing pain	To serve one's own needs or interests, an equal exchange, a deal, an agreement.
	Conventional morality	
3.	What significant others think	The need to be a good person in your own eyes and those of others. Desire to maintain rules and authority that support stereotypical good behaviour.
4.	What society thinks	Right is contributing to society; laws are to be upheld.
	Post-conventional morality	
5.	Social contract or utility and individual rights	Based on overall utility, the greatest good for the greatest number.
6.	Universal ethical princples	Belief as a rational person in the validity of universal moral principles, and a sense of personal commitment to them.

Source: Adapted from Kohlberg (1976)

Kohlberg's Stages of Moral Development

The next concept to be examined is that of morality and moral reasoning, a field that has been dominated by the work of Lawrence Kohlberg. Kohlberg's Stages of Moral Development were originally intended to explain the progressive development of morals in children. Using a hypothetical moral dilemma, based on whether or not a man should steal drugs for his dying wife (see Kohlberg 1980), Kohlberg observed 50 males from the ages of ten to 28. He noted that, given the same scenario, the reasoning that the respondents offered in response to the scenario became increasingly more sophisticated as they grew older. The six stages are summarized in Table 10.1.

Kohlberg's theory has, however, been criticized for a number of reasons. Firstly, it is based only on a sample of 50 males, and Gilligan (1982) suggests that female moral development is different and will diverge from that of males at the post-conventional level, with women having greater emphasis on caring as the highest value. Secondly, Kohlberg's assumption that all cultures will follow the same stages of moral development and that there are universal truths, morals and values has also been challenged. Snarey (1985) suggests that one should expect there to be some cultural nuances and that Kohlberg's existing stages cannot accommodate such differences because, with particular regard to the post-conventional morality,

Table 10.2 Kohlberg's Stages of Moral Development and their application to interpretation

	Stage of moral development	Application to interpretation
	Pre-conventional morality	
1.	Fear of punishment	Interpretation threatens fine for behaviour that breaks the rules.
2.	Maximizing pleasure/minimizing pain	Interpretation communicates personal costs and benefits with the behavioural prescription.
	Conventional morality	
3.	What significant others think	Interpretative appeals must emphasise how behaviour will affect family, friends etc.
4.	What society thinks	Interpretative appeals emphasize good citizenship.
	Post-conventional morality	
5.	Social contract or utility and individual rights	Impacts and consequences of behaviour are described enabling the person to make an educated choice.
6.	Universal ethical princples	Interpretative message communicates how compliance with the behavioural prescription characterises an ethically principled person.

Source: Adapted from Christensen and Dustin (1989)

these stages are based primarily on Western philosophy. Finally, although Kohlberg posits that there is consistency of moral reasoning from one context to another, this has been found not to be the case (Carpendale 2000).

Despite these criticisms the stages of development have been applied to help understand interpretation used in the management of national parks in the United States (Christenson and Dustin 1989). The stages, along with their application to interpretation, are presented in Table 10.2.

The overlap between Kohlberg's stages and typical western philosophical and ethical thought should be noted. For example, stages 1 and 2 relate to the hedonistic motivation of pleasure seeking and pain avoidance, while stage 5 draws on utilitarian reasoned action and stage 6 corresponds with deontological principles: that one should act in a certain manner as it is the universally accepted 'right thing to do'.

Roggenbuck's Application of Communication Strategies

To some extent, Roggenbuck (1992) has simplified these theories and applied them to communication for recreational purposes. According to Roggenbuck (1992) there are three theoretical standpoints for communication:

- *Applied behaviour analysis.* Visitors can be informed of rewards or punishment that will be administered dependent upon their behaviour. However, while sanctions can be useful, they can create negative feelings from recipients (Malloy and Fennell 1998, Carter 2001a).
- *Central route.* Relevant beliefs of visitors are modified through the delivery of substantive messages (reasoned messages). Such communication initiatives can influence attitudes towards a given subject by informing recipients about the consequences of their actions. New or modified beliefs lead to desired changes in behaviour.
- *Peripheral route.* The message source or medium is the key. Sources considered by visitors to be authoritative or powerful may influence behaviour while other messages may be ignored. The credibility of the source has greater effect than the content.

Again, we see the influence of western philosophical thinking: applied behaviour is linked to punishment and rewards (hedonism) and the central route is linked to reasoned communication (utilitarianism). A third, the peripheral route, considers the additional factor of the source of the message being influential. Clearly there is overlap between all three of these theoretical standpoints. For the purposes of this discussion, however, Kohlberg's stages have been used as the basis of a conceptual framework from which to explore the effectiveness of these different ways of communication.

The Kaikoura Case Study

Introduction

The township of Kaikoura is a small coastal community with a population of approximately 2,760, situated on the east coast of New Zealand's South Island. The landscape of Kaikoura is defined by mountains and sea and it is these natural assets that provide the foundation for the tourism industry. The continental shelf is much nearer the coast at Kaikoura than in most parts of New Zealand and, at the Kaikoura Peninsula, two ocean currents converge (McAloon 1998). As a consequence of these geographical features the sea is rich in minerals and plankton, attracting whales, dolphins, fur seals and, in turn, ecotourists. The fur seals haul-out on a rocky peninsula that is easily accessible by foot or by car, and they can often be seen resting on or under vehicles in the car park. This close proximity of seals to humans can cause problems for the seal population and there is concern that the recommended minimum viewing distance of 20 m is often transgressed (Barton et al. 1998). There are anecdotal reports of seals being patted, or of visitors throwing stones at resting seals to get them to move in order to take better photographs. Visitors frequently approach the seals to within a metre or even closer.

Method

Initially, non-participant observation was undertaken at the seal colony and visitors' seal-viewing behaviour was observed. The existing signage and boards were also assessed for their content. Key informants were consulted, including representatives from the Department of Conservation, local government representatives and tour operators based at the seal colony. A survey was then developed using a hypothetical scenario based on seal-viewing behaviour. As Roggenbuck (1992) explains that communication is most useful for unskilled or uninformed, rather than illegal or careless actions, the scenarios were intended to target uninformed behaviour. Six different rationales were given for displaying the desired behaviour. These different messages were based on Kohlberg's Stages of Moral Development and are presented in Table 10.3.

The survey was undertaken at two different locations in New Zealand during February and March 2004. The survey sample was limited, however, as it was difficult to intercept package tourists; some respondents had to be filtered on the basis of their level of English language ability; and domestic tourists were also under-represented due to the time of year (domestic vacations typically being taken in January). Questionnaires were completed by 372 respondents.

Information Boards at the Colonies

There are a number of information boards at the seal colonies, mainly produced by the Department of Conservation. Table 10.4 summarizes the information on these message boards at the time of the research and also suggests how these messages may be related to Kohlberg's Stages of Moral Development.

The first point to note from this summary is the inconsistencies that are presented in the signage. Two signs recommend a viewing distance of 10 m, and two of 5 m. The photograph in Sign 4 implies that visitors might be able to get as close as a metre from the seals. To some extent these inconsistencies may arise from different phases of signage that have been placed at the colony, leaving older versions remaining, and this is understandable as signage is costly to produce and replace. Since the research was undertaken, some new signage has been placed at the peninsula. The point still stands, however, that signage should be consistent. The second observation that can be made from this summary is to note that the content of the information boards can be applied to the theoretical framework provided by Kohlberg's Stages of Moral Development and that a number of stages are used. Information Board 1 uses stage 4 reasoning and makes an appeal to

Table 10.3 Example of a questionnaire used to gauge visitors' responses to different means of communication at Kaikoura seal colony

The Kaikoura Seal Colony
You have just arrived at the seal colony at Kaikoura. The Department of Conservation is trying to stop too many tourists getting too close to the seals. However, in this scenario, you want to get really close to a seal to get a good photograph. Which of the following are likely to influence you to stay at the required distance?

		Not very likely				**Very likely**
A sign saying...		1	2	3	4	5
A	'Please stay 10 metres from the seals. Please respect New Zealand's wildlife'.					
B	'Please stay 10 metres from the seals. It's up to you to do the right thing'.					
C	'Please stay 10 metres from the seals. Seals can bite'.					
D	'Please stay 10 metres from the seals. Approaching closer will make them retreat to the water'.					
E	'Please stay 10 metres from the seals. Approaching the seals can frighten them and their young'.					
F	'Please stay 10 metres from the seals. Don't spoil this experience for other visitors'.					

Note: A = stage 4, B = stage 6, C = stage 1, D = stage 2, E = stage 5, F = stage 3 (of Kohlberg's Stages of Moral Development).

Table 10.4 Summary of information provision at Kaikoura Seal Colony

Type of information	Content of signage	Suggested stage of moral development	Suggested viewing distance
1 Information board	*Many animals and birds make the peninsula their home. Enjoy them, but please respect their space. 5 metres is close enough.*	Stage 4 respect	5 metres
2. Sign	*PLEASE DO NOT DISTURB. Seals resting and may bite if annoyed.*	Stage 5 reason and Stage 1 fear of punishment	10 metres
	Disturbing seals is an offence under the Marine Mammals Protection Act. [Recommends viewing distance of 10 metres.]	Stage 1 fear of punishment	
3. Sign	*Seals near the car park are resting. Do not disturb. Enjoy your visit, but do not attempt to touch or move a seal.*	Stage 5 reason	10 metres
	Seals are wild animals and may bite causing serious injury. [Recommends viewing distance of 10 metres.]	Stage 1 fear of punishment	
4. Interpretation board	*Seals often rest on the rocks near here ... It is OK to take a closer look but keep about five metres between you and the seals – they will let you know if you get too close.*	Stage 5 reason	5 metres (stated)
	It doesn't pay to come between the seals and the sea or to tower over them or move around suddenly. [Interpretation board also shows photograph of a visitor within 1 metre of a seal with caption 'Close eh?']	Stage 1 fear of punishment	1 metre (implied by photograph)

citizenship, to respect the wildlife. Sign 2 provides a limited amount of stage 5 reasoning for not approaching close to the seals – 'seals resting', but the majority of the message relies on stage 1 punishment, both in terms of the seals biting, and in terms of the legal implications. Sign 3, again, provides a limited amount of stage 5 reasoning – 'seals near the carpark are resting', but the majority of the message relies on stage 1 fear of punishment that seals may bite. Interpretation Board 4 also provides a limited amount of information that the seals are resting, but stage 1 punishment is also implied – 'it doesn't pay to come between the seals and the sea' (albeit without making explicit the consequences of this action). This point can be made for the signs that provide some reason combined with punishment; the links between certain behaviours and the effects that this has on the seals are not clearly made. A final point to make is that there is some reiteration of information within the signage. Sign 2 and Sign 3, for example, request that the seals are not disturbed, possibly reinforcing this point as visitors make their way around the site.

Survey Results

This section examines the data from the survey. It begins with the modal response to all of the messages, where respondents were asked to rank each message (and hence Kohlberg stage) on a scale of 1 to 5 according to its effectiveness in influencing their behaviour (5 being very likely to influence behaviour and 1 being not very likely to influence behaviour). Figure 10.1 shows that stages 1, 2, 4 and 5 are all rated as 5 (very likely to influence behaviour). Indeed, many of the respondents ticked 5 for all the stages, explaining that all of the messages would be likely to influence them. However, the modal responses for the stage 2 and stage 6 messages drop to 3, indicating that these messages might be less likely to influence visitor behaviour.

From Figure 10.2, which presents the mean response on a scale of 1 to 5, a clearer picture of preference emerges. The messages based on stages 4 and 5 of development appear to have the greatest influence on visitors. Again, the messages based on stages 3 and 6 of moral development have relatively low means. From Figure 10.1 then, it is possible to see that overall modal responses were high, and from Figure 10.2 it can be seen that stages 4 and 5 were rated as more likely to influence behaviour.

Figure 10.3 highlights those messages identified by the respondents as the most and least likely overall to influence their behaviour. This is particularly important as for the initial part of the question, rating each message on a scale of 1 to 5, many respondents chose 5 (very likely to influence behaviour) for all the messages with little distinction between them.

From Figure 10.3 it can be seen that the message most likely to influence behaviour is based on stage 5 of moral development, which relates to utilitarian reasoning. Forty-one per cent of respondents stated that this message would be the most likely to influence their behaviour. The second most likely message to

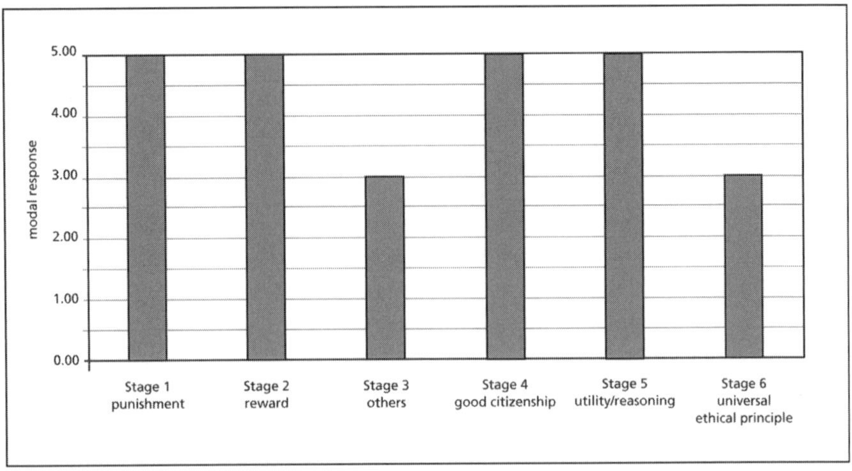

Figure 10.1 Modal responses from the visitor survey

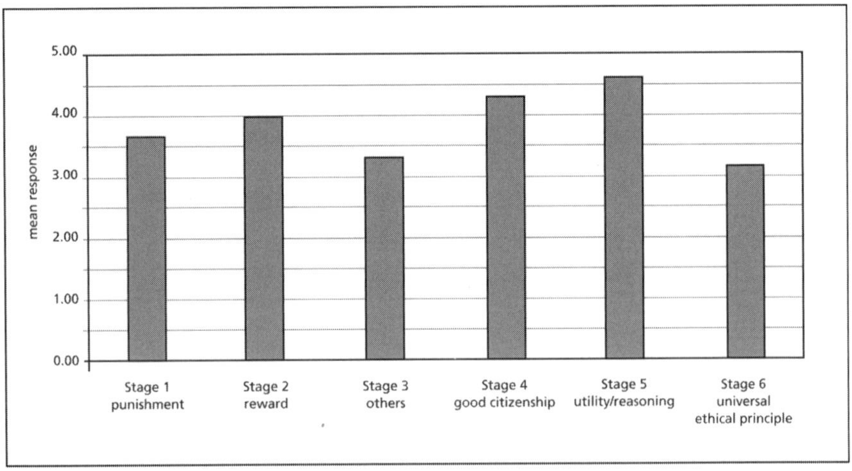

Figure 10.2 Mean responses from the visitor survey

influence behaviour is based on stage 4, relating to good citizenship, with 35 per cent of respondents stating that this would be most likely to influence their behaviour. Messages based on stages 2 and 3 of moral development, reward and considering others, are not particularly influential (7 per cent and 2.5 per cent of visitors chose these messages respectively). Finally, the message based on stage 1, punishment, may have some influence with 12.5 per cent of respondents stating that this would be the most likely message to influence them. The message 'to do the right thing', based on stage 6, universal ethical principles, has little influence

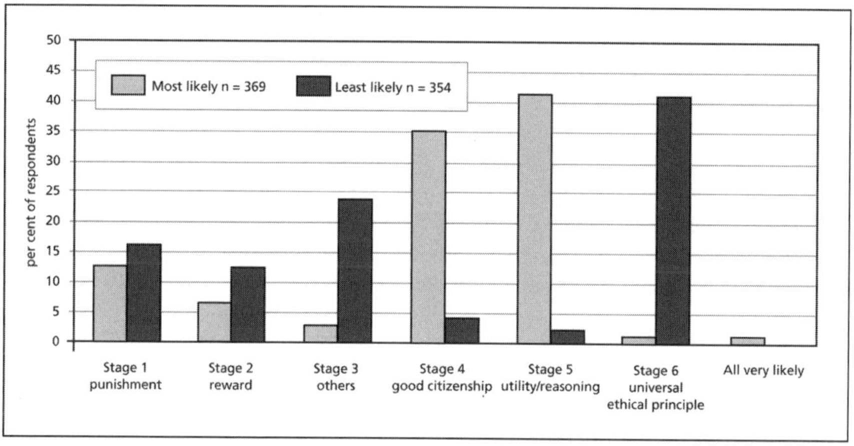

Figure 10.3 Summary of messages most and least likely to influence visitor behaviour

with just 1 per cent of respondents stating this would be the most likely message to influence their behaviour. A small number of respondents did not choose any single message as being the most likely overall to influence them, stating that all messages would be very likely to influence their behaviour.

Respondents also gave their reasoning as to why they chose a certain message as the most likely to influence them. This allowed key reasons to be established as to why messages were influential, based not on how messages were intended by the researcher, but on how they were interpreted by the recipient. Indeed, the stage of the message did not always correspond with the interpretation of the respondents. For example, many respondents chose the stage 5 message (the reasoned argument), but explained their choice in terms of stage 4 development: that they wanted to respect wildlife. It seems that giving an explanation of signs is useful in evoking a response and creating a greater understanding and respect, even if this is not how the message is interpreted.

As far as the least likely messages to influence behaviour are concerned, this is something of a mirror image of the most likely responses. The messages based on stage 6, universal ethical principles, seem overall to be the least likely to influence behaviour. Forty-one per cent of respondents stated that this would be the least likely to influence their behaviour. Messages based on stage 1, punishment; stage 2, reward; and stage 3, considering others, are also unlikely to influence behaviour (17 per cent, 12.5 per cent and 23.5 per cent respectively).

The respondents were again asked to express in their own words why they had ranked a certain message (and hence Kohlberg stage) as least likely to influence their behaviour. Because they are 'negative' is a common response regarding why the stages 1, 2 and 3 messages are the least likely to influence behaviour, with respondents stating that these sorts of messages are patronizing, harsh, threatening, too greedy and so on. Stage 1 and 2 messages are frequently chosen as being the least likely to influence behaviour as the respondents negotiate with the content of the message arguing, for example, that they could outrun any seal intending to bite them. Negotiation also occurs to a lesser extent with the stage 4, 5 and 6 messages. Some respondents state that messages are unlikely to influence their behaviour because they feel that the message is untrue or they can discredit it, and this reason is found particularly for stages 2, 3 and 5. Stage 3 messages, which appeal to the respondent to think of others, frequently evoked a response 'would other people think of me?' Overall, however, the main reason given to explain why a message is unlikely to influence behaviour is because it is 'not reasoned'. This response is given for the messages at all stages, but is most noticeable at stage 6; respondents simply do not know what 'the right thing' is.

Discussion and Conclusions

The results indicate that it is important to know not only which messages are influential, but why they are so. Similarly, it is also helpful to understand which messages are not influential, and why. An effective type of communication would combine this knowledge. The most likely messages to influence behaviour are stage 4 and 5 messages. With respect to respondents' own interpretation of how such messages are understood, appeals to be a good citizen and to demonstrate respect are important. As far as messages that are least likely to influence behaviour are concerned, stage 6 (universal ethical principles) messages lack influence, as insufficient information is provided and respondents do not know what the right thing to do is in this context. Stage 1 (punishment) messages are less influential, typically because of the negative nature of the message. The stage 3 message also has little influence, with many respondents indicating that they did not understand how their inappropriate action would affect others.

Tourists in a novel situation do not intuitively know right from wrong and required behaviour should, therefore, be explained to them in order to be more effective. The importance of reason was further reinforced during data collection, when respondents were being interviewed. Many respondents asked for more information regarding the seals and what was considered appropriate behaviour at the colony. Several respondents stated that they had previously visited the seal colony and had approached closer than the recommended viewing distances, but they would not have done so had they realized that their actions may have caused distress to the animals. Many who had not yet visited indicated that they would be more mindful and considerate during their intended visit due to an increased understanding of potentially harmful visitor behaviour (following Moscardo 1996).

If we revisit the existing signage at the seal colony in the light of this knowledge it is also possible to understand a little further whether or not this information is likely to be effective. Based on the evidence in this context, effective communication would include an appeal to good citizenship, combined with a reasoned and positive argument. This is consistent with research in the area of management ethics based on rule-utilitarian approaches, whereby the individual can learn through an understanding of the consequences of their actions (Malloy and Fennell 1998). However, the signage at the colony, at the time of the research, provides little reasoning and is often negative in tone. In addition, we can note from the research process itself that many respondents were excluded from participating in the survey because they did not speak English as a first or second language (and the survey had not been translated). Evidently, the language of the research and of the information itself is important and cannot be effective if it is not understood. The information at the seal colony is presented only in English and, both from the point of view of better hospitality as well as increasing the effectiveness of messages, it should be provided in different languages.

Kohlberg's stages and wider ethical philosophies, if somewhat simplified here, are a useful framework for establishing effective types of communication designed to influence responsible behaviour. Although creating scenarios and messages that truly reflect the stages of moral development can be challenging, the framework provides a useful theoretical foundation to follow when developing any kind of communication designed to influence behaviour, regardless of the medium of the message, be it interpretative or code of conduct, oral or written.

There are limitations to this work, however. This study was based on a hypothetical scenario, and signage or other forms of communication should be trialled in an experimental situation in the field. The respondents of this survey were a somewhat captive audience, and once they had agreed to the questionnaire they were guaranteed to read the messages. There is, of course, no such guarantee in the real world, and ensuring that the recipients read the message is as important as providing the most effective message. Finally, as stated earlier, the survey relied on respondents understanding the semantic differences of the messages and many potential respondents were excluded because they did not have English as a first or confident second language. Those from other cultures (where English is not typically spoken or learnt as a second language) may react very differently to the stages of moral development as presented here and this invites the need for further, multi-lingual and multi-cultural research. Different contexts should also be considered and similar research could be undertaken that tests this theoretical framework in different situations. Research that investigates the potential effect of reiterated signage would also be worthwhile.

Effective communication is not, however, just about getting the words right. Information also has to be considered with regards to the source or media of the message and also the placement of the message. Consequently, establishing the optimum positioning and mode of communication may be as important as the message itself. It can be seen that there are many media that impart information. These can include formal written signage as well as spoken information. Written information could be from a brochure, guidebook or other marketing, from interpretation and from codes of conduct, and these could be placed in any number of situations. Oral information may be given by tour operators and guides, by accommodation providers, by other tourists and by transport providers. Indeed, many tourists commented on the influence of the information that was given to them by bus drivers, for example, and discussed how this had influenced them. These messages should be reinforced by a wide network of delivery agents, and information providers should be encouraged to supply guidance that is most appropriate for each individual audience.

Acknowledgements

I would like to thank Professor Douglas Pearce and Dr Karen Smith for their input whilst this research was undertaken, as part fulfilment for a PhD thesis. Thanks also to the Commonwealth Scholarships Committee, the Victoria University of Wellington completion scholarship, and the Victoria Management School for funding this work.

References

Barton, K., Booth, K., Ward, J., Simmons, D. and Fairweather, J. (1998) *Visitor and New Zealand fur seal interactions along the Kaikoura coast, Kaikoura case study* (Report No. 9). Tourism Research and Education Centre, Lincoln University, Canterbury.

Boniface, P. (1999) Tourism cultures: consensus in the making? In M. Robinson and P. Boniface (eds) *Tourism and Cultural Conflicts*. CABI Publishing, Wallingford, pp. 287–306.

Boyd, S.W. and Butler, R.W. (1996) Managing ecotourism: an opportunity spectrum approach. *Tourism Management*, 17, 557–566.

Broadhurst, R. (2001) *Managing Environments for Leisure and Recreation*. Routledge, London.

Brody, B. (1983) *Ethics and its Applications*. Harcourt Brace Jovanovich, New York.

Butcher, J. (2003) *The Moralisation of Tourism: Sun, Sand and Saving the World?* Routledge, London.

Carpendale, J.I.M. (2000) Kohlberg and Piaget on stages and moral reasoning. *Developmental Review*, 20, 181–205.

Carter, J. (2001a) A review of research on changing recreational behaviour through communication. In M.B. Usher (ed.) *Enjoyment and Understanding of the Natural Heritage*. The Stationery Office Ltd, Edinburgh, pp. 93–103.

Carter, J. (2001b) Putting out fires with imagination: Using interpretation to encourage desired behaviour. In M.B. Usher (ed.) *Enjoyment and Understanding of the Natural Heritage*. The Stationery Office Ltd, Edinburgh.

Cater, E. and Goodall, B. (1997) Must tourism destroy its resource base? In L. France (ed.) *Sustainable Tourism*. Earthscan, London, pp. 85–89.

Christensen, H. and Dustin, D. (1989) Reaching recreationists at different levels of moral development. *Journal of Park and Recreation Administration*, 7, 72–80.

Eber, S.E. (1992) *Beyond the Green Horizon: A Discussion Paper on Principles for Sustainable Tourism*. WWF UK, Godalming, Surrey.

Forsyth, T. (1996) *Sustainable Tourism: Moving from Theory to Practice*. WWF UK and Tourism Concern, Godalming, Surrey.

France, L. (1997) *Sustainable Tourism*. Earthscan, London.

Gilligan, C. (1982) *In a Different Voice*. Harvard University Press, Cambridge, MA.

Gunn, C. (1988) *Tourism Planning*. Taylor & Francis, New York.

Hall, C.M. and McArthur, S. (1993) Heritage management: an introductory framework. In C.M. Hall and S. McArthur (eds) *Heritage Management in New Zealand and Australia*. Oxford University Press, Oxford, pp. 1–17.

Kohlberg, L. (1976) Moral stages and moralization: the cognitive-developmental approach. In T. Lickona (ed.) *Moral Development and Behaviour: Theory, Research and Social Issues*. Holt, Rinehart and Winston, New York, pp. 31–53.

Kohlberg, L. (1980) Stages of moral development as a basis for moral education. In B. Munsey (ed.) *Moral Development, Moral Education and Kohlberg: Basic Issues in Philosophy, Psychology, Religion and Education*. Religious Education Press, Birmingham, AL, pp. 15–100.

Krippendorf, J. (1984) *The Holiday Makers: Understanding the Impact of Leisure and Travel*. Butterworth Heinemann, Oxford.

Luzar, E.J., Diagne, A., Gan, C.E.C. and Henning, B.R. (1998) Profiling the nature-based tourist: a multinomial logit approach. *Journal of Travel Research*, 37, 48–56.

Malloy, D. and Fennell, D. (1998) Codes of ethics and tourism: an exploratory content analysis. *Tourism Management*, 19, 453–461.

Manfredo, M. (ed.) (1992) *Influencing Human Behaviour: Theory and Application in Recreation, Tourism and Natural Resource Management*. Sagamore Champaign, IL.

Mann, M. (2000) *The Community Tourism Guide: Exciting Holidays for Responsible Travellers*. Earthscan, London.

Manning, R.E. (1999) *Studies in Outdoor Recreation: Search and Research for Satisfaction*. Oregon State University Press, Corvallis, OR.

Mason, P. and Mowforth, M. (1995) *Codes of Conduct in Tourism*. University of Plymouth, Plymouth.

Mason, P. and Mowforth, M. (1996) Codes of conduct in tourism. *Progress in Tourism and Hospitality Research*, 2, 151–167.

McAloon, J. (1998) *Kaikoura: Historical Background*. Kaikoura Case Study, Report No. 1. Tourism Research and Education Centre, Lincoln University, Canterbury.

Moscardo, G. (1996) Mindful visitors: heritage and tourism. *Annals of Tourism Research*, 23, 376–397.

Moscardo, G. (1998) Interpretation and sustainable tourism: functions, examples and principles. *The Journal of Tourism Studies*, 9, 2–13.

Prosser, R. (1992) The ethics of tourism. In D.P. Cooper and J. Palmer (eds) *The Environment in Question: Ethics and Global Issues*. Routledge, London, pp. 37–50.

Reisinger, Y. (1997) Social contact between tourists and hosts of different cultural backgrounds. In L. France (ed.) *Sustainable Tourism*. Earthscan, London, pp. 129–134.

Roggenbuck, J.W. (1992) Use of persuasion to reduce resource impacts and visitor conflicts. In M.J. Manfredo (ed.) *Influencing Human Behaviour: Theory and Application in Recreation, Tourism and Natural Resource Management*. Sagamore, Champaign, pp. 149–208.

Snarey, J. (1985) Cross-cultural universality of social–moral development: a critical review of Kohlbergian research. *Psychological Bulletin*, 97, 202–232.

Thomson, A. (1999) *Critical Reasoning in Ethics: A Practical Introduction*. Routledge, London.

Tilden, F. (1977) *Interpreting our Heritage*. North Carolina Press, Chapel Hill.

Wheeler, B. (1994) Ecotourism: a ruse by any other name. In C.P. Cooper and A. Lockwood (eds) *Progress in Tourism, Recreation and Hospitality Management*. John Wiley & Sons Ltd, Chichester, pp. 3–11.

Chapter 11

Can the Conservation Attitudes and Behavioural Intentions of Tourists to Tropical Forest be Improved through Biodiversity Interpretation? A Case Study from Australia

Jennifer Hill and Georgie Gough

Introduction

Nature-based tourism is one of the fastest growing sectors within the global tourism industry and much of this growth has occurred within fragile and protected environments (Buckley 2000, Mason 2005, Marion and Reed 2007, Weaver and Lawton 2007). For this reason, concern about the environmental impacts of nature-based tourism has risen (Wearing and Neil 1999, Tubb 2003, Kruger 2005) and it has been suggested that its negative impacts can be reduced by the provision of environmental information for tourists (Kimmel 1999). Such ecotourism interpretation acts not only to increase visitor satisfaction and knowledge, but also to encourage attitudinal and behavioural shifts in visitors (Moscardo 1999). This chapter uses published literature to examine the links between interpretation and visitor satisfaction, knowledge, attitudes and behaviour. Further, it presents empirical research from Queensland, Australia, which examines whether educating visitors about biological diversity during a visit to tropical rainforest can raise their levels of satisfaction and knowledge above those achieved from the experiential encounter alone. Selected aspects of biological diversity are interpreted in order to elucidate which are more conducive to enhancing knowledge. Visitor attitudes post-visit are used to test whether increasing satisfaction and/or knowledge can encourage i) more positive attitudes towards ecosystem conservation and ii) more positive behavioural intentions in the ecosystem in future.

There is evidence that on-site rainforest interpretation contributes substantially to tourist satisfaction at the Skyrail Rainforest Cableway in Queensland, Australia (Moscardo and Woods 1998, Pearce and Moscardo 1998). Skyrail visitors who used interpretive facilities were significantly more satisfied with their visit than those who did not, although there appeared to be no account taken of the potential influence of *a priori* visitor attitudes on these results. The existence of such a relationship may be important to tourism management as it has been postulated that when visitors enjoy their visit, they are more likely to support on-site management strategies (Moscardo 1999). Additionally, comparison of visitors to Skyrail with different levels of exposure to the interpretation also demonstrated that the more interpretation the visitors experienced, the more they learned. Similarly, an examination of the impact of interpretive signs on visitor knowledge at the Valley of the Giants Tree Top Walk in Western Australia found significant increases in visitor knowledge and satisfaction as a result of their reading trail-side signs (Hughes and Morrison-Saunders 2002). The study also found a significant positive relationship between the number of visitors claiming a quality learning experience and installation of additional trail-side signs.

It has been claimed widely that interpretation is an important element in the promotion of environmental conservation and sustainable tourism, largely by increasing visitor knowledge and environmental awareness, and by prompting more responsible tourist behaviour (Tilden 1977, Manfredo 1992, Bramwell and Lane 1993, Orams 1994, 1995, Beck and Cable 1998, Moscardo 1999, 2003, Ritchie et al. 2003). Early theorists suggested a simple linear relationship whereby if people became more knowledgeable about the environment, they would in turn become more aware of its problems and thus be more motivated to act responsibly (Hungerford and Volk 1990). Later models identified both location- and person-specific variables which could influence the intent towards and achievement of responsible behaviour (Azjen 1991, Ajzen and Driver 1992, Sparks and Guthrie 1998, White 2005).

The links between interpretation, knowledge, attitudes and behaviour have rarely been tested empirically (Marion and Reed 2007), although a study in Dartmoor National Park in the UK found that increasing visitor knowledge via interpretation did encourage visitors to see how they could behave more respectfully towards the local environment. No significant changes were found, however, in the attitude of post-visit respondents towards the environment more generally (Tubb 2003). More recently, Hughes and Morrison-Saunders (2005a) found that the Tree Top Walk in Western Australia, which offers low intensity interpretation and restricts activities to a passive observation of nature, encouraged a more ecocentric view of the forest by visitors. Also in Australia, Orams (1997) found that participants in the Tangalooma Dolphin Feeding Programme (near Brisbane) reported that this interpretive experience made them more environmentally responsible in their subsequent behaviour, and a similar outcome was identified by Tisdell and Wilson (2005). They explored the role of ecotourism in promoting environmental learning and sustaining nature conservation in Mon Repos Conservation Park in Queensland. They found the ecotourism experiences to be highly effective in terms of environmental education and to have significant positive impacts on the visitors' stated desire and intended behaviour to protect sea turtles. Such impacts on environmental attitudes can, however, depend upon types of visitor activity (Hughes and Morrison-Saunders 2005b). Research on Penguin Island, Western Australia, discovered that exploration-focused visitors shifted towards a conservation-oriented stance immediately post-visit, whilst recreation-focused groups shifted towards a human-centred view of the natural area. Finally, in a different context, Lee and Moscardo (2005) explored changes in tourists' environmental knowledge, awareness, attitudes and behavioural intentions after staying at a major ecotourism resort on Fraser Island in Queensland, Australia. The study suggested that satisfying experiences in ecotourism accommodation and awareness of in-resort environmental practices could reinforce visitors' favourable environmental attitudes.

There has been much less evaluation of tourist understanding of biological diversity (Kerley et al. 2003) and how this can alter visitor attitudes and behaviour. Nevertheless, this is an important issue to address if the potential of biodiversity as an ecotourism resource is to be assessed.

Whilst ecotourism might appear to offer sustainable tourism development, it is duly acknowledged that rainforest ecotourism, with its frequent reliance on long-haul air travel, can be challenged as inherently unsustainable (Mowforth and Munt 2003). Recent analyses have shown, however, that whilst ecotourism does exert an absolute net demand on global natural resources, its global ecological footprint is likely to be considerably less than that of mass tourism and some ecotourism products may potentially make a positive contribution to resource conservation at a global scale (Hunter and Shaw, 2005, 2007 and Chapter 3 herein). The low impact nature of visitor-resource interaction, coupled with altered visitor attitudes beyond the immediate destination area, form part of the off-setting process.

Field Site and Research Methods

Field research was undertaken in the Daintree rainforest in Queensland, Australia (Figure 11.1). This is one of the oldest extant tropical forest ecosystems, which harbours primeval growth forms as a consequence. This forest provides a record of the ecological and evolutionary processes that have shaped the flora and fauna of Australia over the past 415 million years (Wet Tropics Management Authority 1995). Visitor sampling took place at Crocodylus Rainforest Village. The site incorporates a privately owned, unsurfaced, 1.75 km circular ropewalk for which biodiversity information was controlled explicitly by the researchers. The walk was ideal for a study of this kind as it exemplified a diverse range of forest habitats and processes over a relatively small area and it offered one point of access and egress (Figure 11.2). In total, 267 visitors to the ropewalk were interrogated by questionnaire during July/August 2004 and July 2005. Additionally, 34 visitors undertook semi-structured interviews in July 2006 to allow qualitative exploration of the quantitative results. In this chapter, only those responses concerning visitor conservation attitudes and behavioural intentions are examined from the qualitative interviews. As a consequence, the methods below detail the quantitative sampling process.

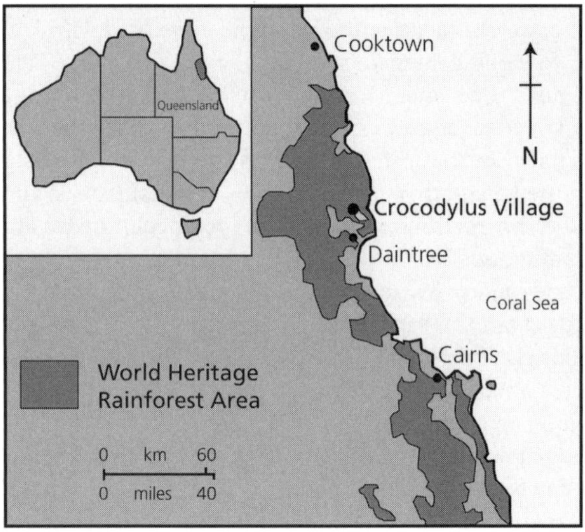

Figure 11.1 Location of Crocodylus Rainforest Village, Queensland, Australia

Visitors undertook a self-administered, written questionnaire and biodiversity quiz. Every visitor who started the ropewalk between 7am and 5pm was asked to take part in the survey and the response rate was 91 per cent. Sampling was divided into two categories of English speaking visitors: those *with* (n=139) and those *without* (n=128) biodiversity information (see below). These categories were self-selecting, although it must be noted that it took more time in both sampling years to amass the required sample of visitors who wished to undertake the walk without information. All visitors undertook the questionnaire immediately pre- and post-visit and they completed the biodiversity quiz post-visit.

Sixteen biodiversity information sheets were used in the research, interpreting aspects ranging from historical evolution of tropical rainforest, through simple facts concerning contemporary rainforest structure, to more complex processes maintaining biological diversity. Rainforest uses, threats and conservation were also outlined. Interpretation was viewed as 'an educational activity which aims to reveal meaning and relationships … by firsthand experience and by illustrative media' (Tilden 1977: 8). The interpretive sheets were A4-sized laminated cards, bound together in the order in which they were encountered along the ropewalk. Visitors were instructed to read the sheets at specified locations along the walk, indicated on a map that formed the cover for the biodiversity information sheets. The positions at which to read the sheets were indicated in the field by prominent numbered markers. As visitors represented a non-captive audience (Ham 1992), they were asked to read only as many of the sheets as they desired during their walk. In post-visit questioning, 77 per cent indicated that they read all sheets and 12 per cent said that they omitted only one or two sheets. The main reasons cited

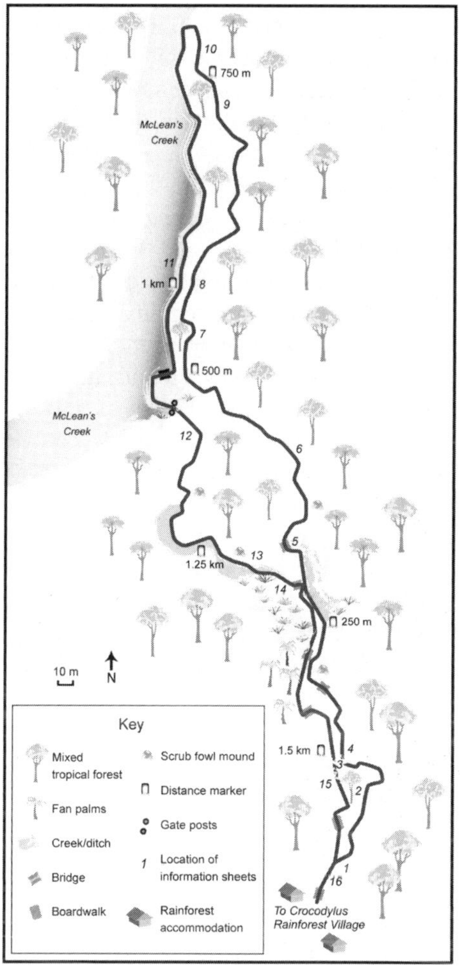

Figure 11.2 The circular rope walk at Crocodylus Rainforest Village, Queensland, Australia, showing numbered biodiversity information points

for omitting sheets were missing the numbered markers and being distracted by environmental conditions (particularly watching footing on the muddy track and looking out for wild pigs). The style and content of the sheets followed key elements of effective interpretation, particularly the inclusion of colour and illustrations; use of grabber headlines and small quantities of text; adoption of a uniform style; and layering of content complexity (Moscardo 1996a).

Questionnaires comprised five sections for those interviewed with biodiversity information, and four sections for those interviewed without information. The sections consisted of trip characteristics; pre-visit survey questions (concerning visitor expectations, motivations, prior rainforest experience, and knowledge and initial conservation attitudes); post-visit general questions (concerning visitor satisfaction, subjective learning and 'return' conservation attitudes); post-visit questions for those receiving biodiversity information (concerning visitor interaction with the information); and visitor socio-demographic characteristics. Visitors were monitored as they answered pre-visit questions to ensure that they did not read through the remainder of the questionnaire. This reduced bias in their answers through pre-visit sensitization (Lee and Balchin 1995). Respondents were not informed about the biodiversity quiz so they did not read the interpretive sheets with the aim of recalling information for a cognitive test.

Motivation of visitors was ascertained by asking how important 15 statements were to their rainforest visit. Visitors responded using a Likert-type self-avowal scale with answers graduated to four possibilities: 1 was 'not at all important', 2 was 'slightly important', 3 was 'important' and 4 was 'very important'. This method provides a snapshot of the multiple motivations of visitors (Holden and Sparrowhawk 2002). Visitor satisfaction with 16 components of their rainforest experience was measured on a scale from 1 to 5, where 1 was 'very dissatisfied', 2 was 'dissatisfied', 3 was 'indifferent', 4 was 'satisfied' and 5 was 'very satisfied'. Questions about whether the rainforest walk met overall visitor expectations and how much visitors thought they had learnt from experiencing the rainforest were established as semantic differentials. Such question styles were adopted as it has been shown that respondents prefer verbal to numerical labels and a verbal anchor reduces over-use of scale extremes (Haley and Case 1979). Some questions were open-ended to explain the quantitative visitor responses. Chi-square (χ^2) and Student t (t) statistical analyses were used to test for significant differences in levels of visitor satisfaction and knowledge according to exposure to biodiversity information. They were undertaken using SPSS version 15 for Windows.

The biodiversity quiz tested visitor cognition post-visit at two levels: recall of factual knowledge (e.g. satinash tree species are good examples of cauliflory) and evidence of deeper understanding (e.g. when forest fragments are far apart, there is little exchange of genetic material between them). Visitors responded to 32 statements, based directly on information provided on the interpretive sheets, by ticking 'true', 'false' or 'don't know' options. Respondents were also asked to tick a box for each question if they could have answered the question prior to their visit. This was used to gauge the level of newly acquired rather than prior knowledge. The resultant scores provided an objective evaluation of short-term visitor cognition.

Results

Survey Sample Demographics and Motivations

The total survey sample of 267 respondents was comprised largely of visitors from outside of Australia (65 per cent). Of the 35 per cent visiting from within Australia, just 3 per cent were from within Queensland. Most of the international visitors were from the UK and Ireland (27 per cent) or the United States and Canada (17 per cent). The highest proportion of visits was made by couples (38 per cent), by groups of friends (22 per cent), or by individual visitors (18 per cent). Families comprised 16 per cent of the sample, whilst tour group participants made up only 6 per cent. Almost 90 per cent of respondents were first time visitors and, of those who had visited the ropewalk previously, their proportion was biased towards a single visit (63 per cent) compared to multiple visits (37 per cent). Only 28 per cent of respondents, however, had never visited rainforest prior to their site visit, 34 per cent had visited rainforest within Australia, 23 per cent had visited a single rainforest location outside of Australia and 15 per cent had visited a number of rainforests around the world.

The majority of the respondents (44 per cent) were aged between 21 and 30. Seventeen per cent and 14 per cent of respondents comprised the 31–40 and 41–50 age groups respectively, and approximately 10 per cent of respondents each were younger than 21 or between 51 and 60 years of age. A small proportion of respondents (4 per cent) were aged over 60 years. There was a slight gender bias in the survey sample with 58 per cent of the sample being female and 42 per cent being male. The sample group was well educated, with 34 per cent having school or college certificates, 31 per cent possessing an undergraduate degree and 35 per cent possessing a graduate degree.

Chi-square tests were undertaken to investigate if there were any significant differences in the socio-demographic profiles for those visitors *receiving* and *not receiving* biodiversity information. The tests showed no significant differences between the two samples (at p=0.05). It can be concluded, therefore, that any significant differences in levels of visitor satisfaction, knowledge and attitudes according to the provision of biodiversity information should not be explained by the varying demographic profiles of the two samples.

Considering all respondents, the strongest motives for visiting the rainforest were to encounter scenic beauty (with a mean response of 3.66), to see the rainforest wildlife (3.61) and to enjoy the sounds, smells and feel of the rainforest (3.56) (Table 11.1). These reasons were considered to be important/very important in motivating the visitors to undertake their walk. The only other motives rated above 3.0 were exploring something new and different (3.40), having fun/enjoyment (3.36) and encountering the rainforest structure and trees (3.22). Learning about the rainforest was the seventh most important motive, returning a mean value of 2.97. The only motive to return a mean value of less than 2.0 was passing time whilst in Queensland (1.83). This reason is, therefore, only slightly important in

Table 11.1 Mean motivation scores for undertaking the rainforest walk, based on 15 statements

Motivation statement	Mean score
Encountering scenic beauty	3.66
Seeing the rainforest wildlife	3.61
Enjoying the sounds, smells and feel of the rainforest	3.56
Exploring something new and different	3.40
Having fun and enjoying yourself	3.36
Encountering the rainforest structure and trees	3.22
Learning about the rainforest	2.97
Escaping from the stress of everyday life	2.79
Taking exercise	2.63
Being creative such as painting or taking photos	2.54
Doing something to tell friends about	2.50
Gaining a sense of rainforest history	2.44
Sharing your knowledge with others	2.40
Being with friends and family	2.35
Passing time whilst in Queensland	1.83

Note: 1 is not at all important, 2 is slightly important, 3 is important and 4 is very important.

motivating visitors to undertake the rainforest walk. Only one motive returned a significantly different result when comparing those visitors with and without biodiversity information. Escaping the stress of everyday life was significantly more important as a motive for those visitors who chose to undertake the walk *without* information (2.97 versus 2.63) ($t=-2.82$, $p<0.01$).

Visitor Satisfaction and Knowledge

Visitors rated their overall satisfaction with the rainforest walk and with 16 components of their rainforest experience. In general, a high degree of visitor satisfaction was expressed irrespective of whether visitors had been provided with interpretive biodiversity sheets. When asked whether the rainforest walk had met their expectations nearly 90 per cent of all visitors agreed/strongly agreed that it had. When all respondents were considered together the highest mean satisfaction score (4.32) was recorded for visitors' encounters with the rainforest structure and trees (Table 11.2). Encountering scenic beauty (4.25), enjoying the sounds, smells and feel of the rainforest (4.16), exploring something new and different (4.15) and having fun/enjoyment (4.10) were the only other categories rated above 4.0. This indicates that visitors were satisfied/very satisfied with these aspects of their rainforest walk. The lowest mean satisfaction scores were recorded for acquiring a sense of rainforest history (2.88), sharing knowledge with family and friends (3.25) and seeing the rainforest wildlife (3.31). Cognitive aspects of learning about

Table 11.2 Visitor satisfaction with the rainforest walk, based on 16 components of their rainforest experience

Satisfaction component	Mean score
Encountering the rainforest structure and trees	4.32
Encountering scenic beauty	4.25
Enjoying the sounds, smells and feel of the rainforest	4.16
Exploring something new and different	4.15
Having fun and enjoying yourself	4.10
Escaping from the stress of everyday life	3.99
Doing something to tell friends about	3.74
Taking exercise	3.72
Learning about the rainforest	3.57
Being with friends and family	3.54
Passing time whilst in Queensland	3.54
Provision of rainforest information	3.42
Being creative such as painting or taking photos	3.34
Seeing the rainforest wildlife	3.31
Sharing your knowledge with others	3.25
Gaining a sense of rainforest history	2.88

Note: 1 is very dissatisfied, 2 is dissatisfied, 3 is indifferent, 4 is satisfied and 5 is very satisfied.

the rainforest (3.57) and provision of rainforest information (3.42) showed largely indifferent visitor responses.

Five components recorded significant differences in visitor satisfaction levels when comparing responses for those visitors receiving and those visitors not receiving biodiversity information (Figure 11.3). Four of these components related to the provision of information, learning and sharing knowledge, whilst one related to experiencing the forest sensually. The most significant difference was recorded for satisfaction with the provision of biodiversity information. Satisfaction was significantly higher for those visitors who used the information sheets (χ^2=89.25, $p<0.001$). Eighty per cent of these respondents said that they were satisfied or very satisfied with the provision of rainforest information, compared to only 23 per cent of respondents who did not use the information. In line with this, mean satisfaction scores rose from 2.75 (dissatisfied/indifferent) to 4.02 (satisfied). Visitor satisfaction in terms of learning about the rainforest was also significantly higher for those visitors who used the rainforest biodiversity information sheets (χ^2=12.13, $p<0.05$). Eighty-two per cent of these respondents said that they were satisfied or very satisfied with their learning about the rainforest, compared to only 32 per cent of respondents who did not use the information. This was reflected in the mean satisfaction scores which rose from 3.04 (indifferent) for the sample without information to 4.04 (satisfied) for the sample with information. Satisfaction gained from sharing knowledge with family and friends was significantly higher for those

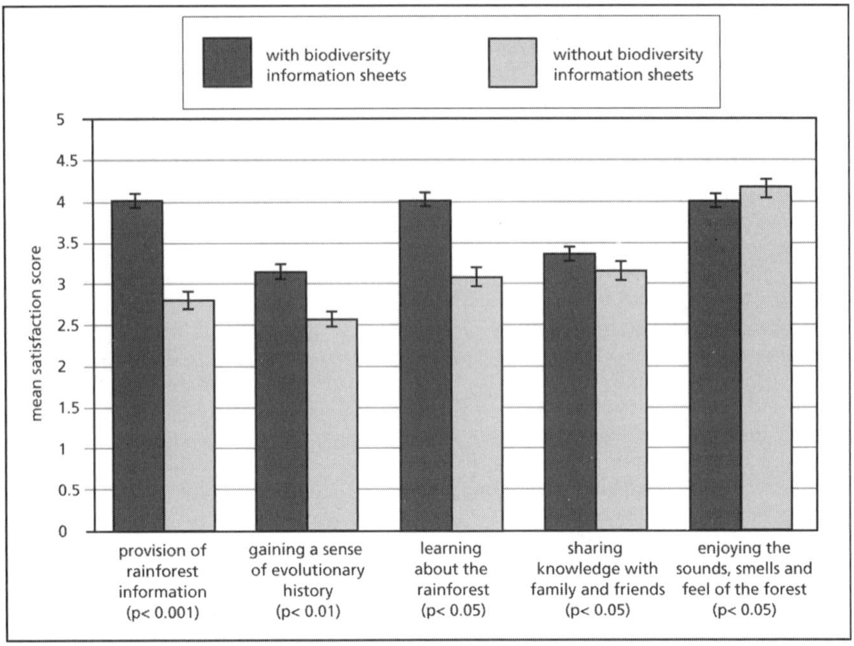

Figure 11.3 Components of the visitor experience for which significant differences were recorded in mean satisfaction scores (± SE) for visitors with and without biodiversity information sheets

visitors who used the rainforest biodiversity information sheets (χ^2=12.44, $p<0.05$). Forty-four per cent of the respondents who used the interpretive sheets on the walk said that they were satisfied or very satisfied in terms of sharing their knowledge with others, compared to 31 per cent of respondents who did not use the sheets. In relation to comprehending something of the evolutionary history of these forests, the satisfaction level of visitors who used the biodiversity information sheets was significantly higher than those who did not use the sheets (χ^2=12.97, $p<0.01$). Mean satisfaction scores rose from 2.60 (dissatisfied/indifferent) to 3.11 (indifferent/satisfied). The final significant difference in satisfaction levels was related to the nature of the visitor encounter with the forest. Seventy-seven per cent of visitors who used the information sheets said that they were satisfied/very satisfied with enjoying the sounds, smells and feel of the forest compared to 89 per cent of those visitors who did not use the sheets. The mean satisfaction score rose from 4.07 for those visitors using the sheets to 4.25 for the visitor sample not using the sheets (χ^2=12.13, $p<0.05$). It must be noted, however, that figures above 4.0 indicate a high degree of satisfaction for both groups of visitors, despite there being a significant difference recorded.

In terms of subjective (perceived) visitor knowledge, 67 per cent of visitors who had received biodiversity information thought that they had learnt a moderate amount/a lot from their rainforest walk, compared to only 38 per cent of visitors who completed the walk without information sheets. Visitors who had received biodiversity information rated their learning experience significantly higher than those who did not receive information (χ^2=28.09, $p<0.001$). This difference derives from a common base of perceived knowledge: 86 per cent of visitors who received information rated their rainforest knowledge as nothing/a little prior to their site visit, compared to 84 per cent of those visitors who did not receive information. Overall, the perceived knowledge profiles of visitors prior to their undertaking the walk was not significantly different between the two samples (at $p=0.05$).

In terms of objective (actual) visitor knowledge, the mean quiz score for the sample of visitors who did not take the biodiversity information sheets on their walk was 40 per cent. This contrasted with a mean score of 70 per cent for the sample of visitors who undertook the walk with the sheets. Importantly, this difference in knowledge was significant ($t=13.93$, $p<0.001$). The sample of visitors who used the biodiversity information sheets gained significantly higher results in the biodiversity quiz. Thus, after being supplied with information visitors believed that they knew more about the rainforest and they had significantly increased their short-term knowledge about the ecosystem, its threats and its conservation. It can also be inferred from the quiz responses that the learning was largely new learning that took place on site, rather than recovery of information learnt prior to the site visit. There was no significant difference, for example, in correct quiz answers for a statement that the cassowary is Australia's biggest rainforest bird. Ninety-two per cent of those visitors who received information acknowledged that this statement was correct, compared to 83 per cent of those who did not receive information. This is a common piece of information that is repeated throughout the Daintree, and the majority of visitors indicated that they could have answered this question prior to their visit. By contrast, an incorrect statement that cauliflory (where flowers and fruit grow on the trunks and major branches of trees) describes the harvesting of produce from the forest floor, drew significantly more correct answers from those visitors who had undertaken the walk with the biodiversity information, compared to those who had not received the information (75 per cent compared to 10 per cent) (χ^2=132.3, p=<0.001). This is a less common and more scientific piece of factual information, so it would be expected that visitors using the information sheets would fail to know the answer to this question prior to their visit, but they should be able to recall the fact post-visit. By comparison, 66 per cent of visitors who had used the biodiversity information sheets correctly identified that when forest fragments are far apart there would be less exchange of genetic material between them, compared to 29 per cent of visitors who had not received the information sheets. The numerical difference between the percentages for right and wrong answers was lower for this question concerning a more complex ecosystem process, but it nevertheless indicated significantly higher knowledge for those receiving the information compared to those who did not (χ^2=53.2, $p<0.001$).

Visitor Attitudes Towards Ecosystem Conservation

All visitors sampled experienced the rainforest ecosystem and half were provided with biodiversity interpretation that informed them of the nature of the ecosystem and its threats. Despite significantly higher cognitive performances by visitors using the sheets compared to those who did not, there was no notable difference in attitudes towards rainforest conservation. Seventy-six per cent of visitors who received biodiversity information and 73 per cent of visitors who did not receive the information said that the rainforest walk had not changed, or changed only a little, their attitude towards rainforest conservation and this difference was not significant (at p=0.05). The qualitative interviews demonstrated that visitors felt they were already conservation-oriented, with one respondent stating 'I was already strongly convinced that it [the rainforest] should be strongly protected'. Another commented 'I'd like to think that I had a positive attitude towards it anyway'. Respondents using the information sheets often referred to a reinforcement or deepening of conservation attitudes rather than an alteration in disposition. One visitor noted, for example, 'not a change in direction … [but] a deeper appreciation', and another commented 'I think [the walk] reinforced rather than changed [my attitude]'. There were, however, some very positive conservation-oriented comments, such as 'I think that before the walk I didn't appreciate what we have and now it makes me want to fight for it' and 'I'll take it [the forest] less for granted'.

Some visitors indicated a role for the interpretive information in altering their conservation awareness directly:

> The logging and how that can affect an entire rain forest ecosystem, I didn't think it was all that sensitive, but now I understand that it is more sensitive that what I even expected. So, yeah, that definitely increased my awareness of the rain forest conservation.

A further comment was that 'I feel more informed about my reasons for thinking that it's important to conserve it [the forest]'. Perhaps of greater import, however, is evidence that the *in situ* learning experience led to a desire to disseminate the conservation message:

> I'd love for my family to come and see this so that they can share what I feel in there and understand and be willing to spread that knowledge and educate other people to the importance of keeping our rain forests and our habitats [from] not being ruined.

Behavioural Intentions of Tourists in Tropical Forest

Visitors to the site did not notably alter their intention in terms of the way they would behave as tourists in the rainforest in the future. Seventy-four per cent of visitors who received biodiversity information and 77 per cent of visitors who did not receive the information said that the rainforest walk had not changed, or changed only a little, their expected behaviour. This difference was not significant (at p=0.05). The interview responses, once again, showed that visitors felt they already behaved as environmentally responsible tourists. One respondent commented 'I think that I am already an 'ethical' walker, as far as staying on the path, not touching or breaking things and so on, but it will make me more observant'. Another noted:

> I think you have to respect any system that you go into, be it a rain forest, be it a temperate forest, be it the African bush. Respect and knowledge is probably the key to sustaining systems like these. So I don't think it really adjusted my perception, it just maybe reiterated the importance of careful conservation.

There was evidence that undertaking the walk with interpretive information could raise visitor self-confidence in the environment, leading to a desire to increase the length of encounter and provoking greater interest and thought: 'I may be a bit more confident to just walk and see what's around', 'if you know nothing it's always the same, but when you know something more you have different views for everything', and '[I would] look at different things and be a little more curious about different parts of the forest that I'm looking at'. This is a responsive change in behaviour towards the environment as a consequence of experiencing it, and it is also prospective, implying future changes in activity within that ecosystem.

Two sheets, entitled 'Clear Today, Gone Tomorrow' and 'Rainforest Uses', were included specifically to educate visitors about the importance of tropical forest. These were picked out by a number of respondents in terms of making them consider the threats to the ecosystem and the need to protect it into the future. One visitor commented 'the more I see the more I have to think about the future'. A number of visitors noted the importance of conserving large areas of forest and reducing even small spatial disturbances such as road construction. Additionally, there was a sense of responsibility for the environment shown by some respondents after their walk:

> I know that the rain forest, the wet tropics, are World Heritage and you have to be careful. It is, of course, if you see it for the first time with your own eyes, you see the amazing environment and feel responsible for it.

There was also a desire shown to teach others about the environment: 'I can imagine … maybe, when I walk with other people, I can point to a few things I learnt and saw on this map. Maybe I can teach other people'.

In answer to a broader question of whether undertaking the walk with the interpretive information had changed visitor environmental attitudes generally, there were some interesting connections made between rainforest conservation and wider environmental issues. One respondent from Melbourne noted:

> At the moment we're in drought, which is having major impacts on our environment and our local areas, so definitely, like with water conservation … and like just how people are being selfish and silly and just taking it for granted and seeing what damage is done.

Another respondent noted that, in terms of altering environmental attitudes, information transfer in itself is simply the start of a more important process where such information must be utilized in discussion: 'It's about sheets providing information, but really the process to get to another attitude is mostly about conversation'.

Discussion

The circular ropewalk at Crocodylus Rainforest Village provided a satisfying tourist experience irrespective of whether visitors had been provided with interpretive biodiversity sheets. Tourists seemed to partition their consumption of the rainforest into an affective domain (subjective emotions and feelings) and a cognitive domain (objective knowledge acquisition). Similar to previous findings (Wallace and Pierce 1996, Ryan et al. 2000, Curtin 2005) visitors were satisfied with gaining a purely affective experience, where the value of their visit lay in the pleasure they gained from a sensual and immersive encounter with an environment that was often novel to them. Key motivations for undertaking the rainforest walk were encountering scenic beauty, experiencing the rainforest sensually, exploring a novel environment and having fun/enjoyment (following Wight 1996, Epler Wood 2002, Niefer et al. 2002 and others). All visitors were satisfied with these aspects of their visit on completion of the walk. This is most likely explained by the nature of the site. The narrow, unsurfaced path with tree trunks and boughs to climb over or pass under provided an immediate and authentic encounter with the natural environment (Figure 11.4). There was incongruence, however, between motivation and satisfaction with regard to seeing the rainforest wildlife. Viewing wildlife was cited as a key reason to undertake the rainforest walk and yet visitors were rather indifferent to the wildlife experience after their visit. They noted a general inability to see animals, whether bird life, reptiles, or the flagship species of the cassowary. Some did question whether this was due to the threats identified on the interpretive sheets reducing population sizes and hence making wildlife spotting difficult. The sheets thereby prompted concern over the resilience and fragility of the forest (see below).

Interpretation did have a significant impact on the emotive encounter of visitors with nature. There was some evidence of visitors being removed from the immediate environment as their cognitive skills were engaged (Cole et al. 1997). This was despite the fact that the sheets were written specifically to draw visitors into the forest via the use of sensory activity boxes. Such activities included, for instance, 'can you see any blue cassowary plums around or on the numbered tree?',

Figure 11.4 A section of the circular ropewalk at Crocodylus Rainforest Village, showing the close encounter of visitors with the rainforest

'count the number of pioneer species on the floor. Use their different leaf shapes to guide you' 'feel the bumpy bark ... this is where the flower buds come from' 'listen for the [pig] grunts'. Overall, however, a high degree of satisfaction with the emotive encounter was recorded for *both* groups of visitors, possibly because learning about the rainforest was cited as an important reason to undertake the walk. In addition, the nature and content of the sheets seemed consistent with the visitors' mode of interaction with the environment; largely appreciative–symbolic, with a focus on exploration, rendering them receptive to on-site interpretation compared to visitors participating in more active–expressive recreational activities (Ballantyne et al. 1998, Hughes and Morrison-Saunders 2005b). These findings agree with a number of studies that nature-based tourists want to satisfy motives of experiencing and learning about their environments (Eagles and Cascagnette 1995, Saleh and Karwacki 1996, Hatch 1998).

It must also be noted that the samples of visitors either using or refraining from using information were self-selecting and that escaping the stress of everyday life was a significantly more important motive for those visitors *not* using information compared to those visitors who chose to take information on their walk. The sample of visitors who chose not to use the biodiversity information opted for no prescriptions on their activities in the forest, whereas the sample of visitors who chose to use the sheets were more willing to accept a mediated encounter with the environment, one that incorporated prescribed stops and interactions with their surroundings in addition to those undertaken under the visitors' own volition. Despite the fact that such 'fixed' information provides relatively few opportunities for visitors to personalize their experience, they were satisfied with their rainforest encounter overall. At times, questions arose that were interesting and relevant personally to visitors, but which could not be answered until the end of the walk via interaction with the interview team. The utility of dynamic interactions with information/informants is supported by research in other contexts (Moscardo 1996b, Moscardo and Woods 1998).

Visitors were dissatisfied with the cognitive experience if they had not used biodiversity information sheets. There were significant increases in visitor satisfaction after using the sheets in terms of learning about the rainforest, particularly its historical evolution, and with the provision of information. In contradiction to Ryan et al. (2000) the cognitive domain complemented the affective domain as visitors without information requested interpretation on their return. This study shows that biodiversity information *per se* was of interest to visitors and it supports the contention that visitors to nature-based attractions without interpretative opportunities can desire cognitive stimulation through information provision (Stein and Lee 1995, Lück 2003). Additionally, the satisfaction gained from sharing knowledge with friends and family was higher for those visitors who used information sheets during their walk. These visitors seemed to enjoy sharing knowledge, the sheets prompting not just self-learning but transmission of knowledge in the form of *in situ* teaching on site. There was less satisfaction gained from sharing observations or knowledge unaided by interpretive information.

Visitors who used biodiversity information sheets during their walk rated their subjective (perceived) learning significantly higher than those who received no information. Exploration of the rainforest without interpretation did not constitute a perceived learning opportunity for the majority, although a minority rated their perceived learning as satisfactory through the nature experience alone (following Stein et al. 2003). The belief that learning has occurred through interpretation (or indeed without it) can be as useful to tourism managers as the objective act of learning, in terms of increasing the quality of the tourist experience (Madin and Fenton 2004).

Biodiversity interpretation played a significant pedagogic role at the site in that it acted as an effective tourist educator. There was a significant difference in objective (actual) cognition between the two groups of respondents. The sample of visitors who used the biodiversity information sheets gained significantly higher results in a biodiversity quiz than those visitors who undertook the walk without the information sheets. Additionally, the short-term learning was largely new learning that took place on site, rather than recovery of information learnt prior to the site visit. It is acknowledged here that, due to the specific aims of the project, the aspects of cognitive development under investigation were limited to Western, science-based ecological thinking (Staiff et al. 2002).

From a tourism perspective, therefore, what are the benefits of providing visitors with an objective learning experience? Whilst visitors to the site exemplified the consumption of nature made both emotional and intellectual, the emotional response to the ecosystem was a partial response. Visitors who used the biodiversity information sheets received a more holistic experience. They rated the sense of rainforest history significantly higher than visitors without information and they also noted the importance of synergistic interactions in maintaining ecosystem functioning. When asked to cite the most important fact that they had learnt from their visit to the rainforest one respondent answered, 'the inter-relations between the plants and animals. If you take one thing out it can change the whole nature of the forest'.

The 'gently provocative' (Bramwell and Lane 1993: 79) provision of biodiversity information did not seem to lead to negative visitor experiences due to over-interpretation. In this way, there is potential for rainforest sites to provide a more rounded satisfaction for visitors by appealing to their cognitive as well as their affective domain. The visitors to this site displayed both intelligence and interest and they can also be described as 'mindful' (Moscardo 1996b, 1999). They consequently deserve quality interpretation. Visitors can be sensitized to information about biological diversity, but without this sensitization they can under-appreciate biodiversity. Ecosystem biodiversity should be valued as an ecotourism resource and its importance conveyed to visitors.

There was no support that on-site interpretation of biological diversity, including pertinent threats and conservation issues, altered visitor attitudes towards sustainable tourism and ecosystem conservation, despite greater cognitive performance by these visitors. This agrees with the work of Beaumont (1998, 2001) who found that environmental attitudes and behaviours did not increase correspondingly with rising environmental knowledge for eco-lodge visitors to Lamington National Park, Australia. Working atop a high self-reported sensitivity to the environment, biodiversity interpretation did not enhance the likelihood that visitors to the rainforest would care more for, or behave differently in, that environment, but it did seem to reinforce visitors' already favourable conservation attitudes and behavioural intentions (following Lee and Moscardo, 2005). A 'ceiling effect' (Beaumont 2001: 335) seemed to have been reached, whereby visitors reported reasonably strong environmental attitudes prior to undertaking the rainforest walk. The interpretive information did, however, seem to mediate the encounter with nature, leading to an expression of desire to gain a longer and more perceptive encounter with the environment in the future. Interpretation seemed to stimulate visitors to engage in a positive way with the rainforest. The reinforcement of conservation attitudes and responsible behaviour within the ecosystem; the recognition of personal responsibility for the protection of the environment and the reduction of its threats; and the desire to teach others about such issues were positive findings. Some respondents were even prompted to consider the nature of attitudinal change. Rather than accepting a direct relationship between education and attitude, they began to consider and articulate a more complex process of attitudinal shift based in discursive reflection. It must also be borne in mind that measurement of conservation attitudes immediately after a site experience does not provide any indication of how these attitudes might alter in the long term (Hughes and Morrison-Saunders 2005b). The reinforcement of positive attitudes here might possibly work with subsequent experiences to shift attitudes and behaviour in future.

To conclude, the results presented here should prove useful to site managers in their quest to provide both a satisfying tourist experience and one that supports conservation of the ecosystem that tourists have come to see. Environmental interpretation and visitor education might offer a 'soft' strategy for managing tourism that complements many 'hard' physical and regulatory options, thereby allowing a more holistic approach to management (Kuo 2002, Mason 2005, Marion and Reed 2007). Whether such environmental education can support a shift from a 'minimalist' mode of ecotourism (superficial learning opportunities and site-specific sustainability objectives) to a 'comprehensive' mode (fostering deep understanding and transformation of behaviour) (Weaver 2005) remains to be seen.

Acknowledgements

The authors would like to thank the owners (Terry and Ian Rogers) and management of Crocodylus Rainforest Village for allowing removal of their usual ropewalk interpretation sheets and replacement with the research project biodiversity sheets. The authors also gratefully acknowledge the kind permission to undertake visitor questioning. Special thanks are extended to Rosemary Burton and Anne Morrison for their enthusiastic support and company in the field. Research funding was provided by the School of the Built and Natural Environment, University of the West of England, Bristol, UK.

References

Azjen, I. (1991) The theory of planned behaviour. *Organizational Behaviour and Human Decision Processes*, 50, 197–211.

Ajzen, I. and Driver, B.L. (1992) Application of the theory of planned behaviour to leisure choice. *Journal of Leisure Research*, 24, 207–224.

Ballantyne, R., Packer, J. and Beckman, E. (1998) Targeted interpretation: exploring relationships among visitors' motivations, activities, attitudes, information needs and preferences. *Journal of Tourism Studies*, 9, 14–25.

Beaumont, N. (1998) Promoting pro-environmental attitudes and behaviours through ecotourism: that's the theory but does it happen in practice? In S. McArthur and B. Weir (eds) *Australia's Ecotourism Industry: A Snapshot in 1998*. Ecotourism Association of Australia, Brisbane, pp. 26–27.

Beaumont, N. (2001) Ecotourism and the conservation ethic: recruiting the uninitiated or preaching to the converted? *Journal of Sustainable Tourism*, 9, 317–341.

Beck, L. and Cable, T. (1998) *Interpretation for the 21st Century*. Sagamore, Champaign, IL.

Bramwell, B. and Lane, B. (1993) Interpretation and sustainable tourism: the potential and the pitfalls. *Journal of Sustainable Tourism*, 1, 71–80.

Buckley, R. (2000) Tourism in the most fragile environments. *Tourism Recreation Research*, 25, 31–40.

Cole, D.N., Hammond, T.P. and McCool, S.F. (1997) Information quantity and communication effectiveness: low-impact messages on wilderness trail-side bulletin boards. *Leisure Sciences*, 19, 59–72.

Curtin, S. (2005) Nature, wild animals and tourism: an experiential view. *Journal of Ecotourism*, 4, 1 15.

Eagles, P.F.J. and Cascagnette, J.W. (1995) Canadian ecotourists: who are they? *Tourism Recreation Research*, 20, 22–28.

Epler Wood, M. (2002) *Ecotourism: Principles, Practices and Policies for Sustainability*. International Ecotourism Society, United Nations Publication.

Haley, R.I. and Case, P.B. (1979) Testing thirteen attitude scales for agreement and brand discrimination. *Journal of Marketing*, 43, 20–32.

Ham, S.H. (1992) *Environmental Interpretation: A Practical Guide for People with Big Ideas and Small Budgets*. North American Press, Golden, CO.

Hatch, D. (1998) Understanding the Australian nature-based tourism market. In S. McArthur and B. Weir (eds) *Australia's Ecotourism Industry: A Snapshot in 1998*. Ecotourism Association of Australia, Brisbane, pp. 1–5.

Holden, A. and Sparrowhawk, J. (2002) Understanding the motivations of ecotourists: the case of trekkers in Annapurna, Nepal. *International Journal of Tourism Research*, 4, 435–446.

Hughes, M. and Morrison-Saunders, A. (2002) Impact of trail-side interpretive signs on visitor knowledge. *Journal of Ecotourism*, 2/3, 122–132.

Hughes, M. and Morrison-Saunders, A. (2005a) Influence of on-site interpretation intensity on visitors to natural areas. *Journal of Ecotourism*, 4, 161–177.

Hughes, M. and Morrison-Saunders, A. (2005b) Interpretation, activity participation, and environmental attitudes of visitors to Penguin Island, Western Australia. *Society and Natural Resources*, 18, 611–624.

Hungerford, H.R. and Volk, T.L. (1990) Changing learner behaviour through environmental education. *Journal of Environmental Education*, 21, 8–21.

Hunter, C. and Shaw, J. (2005) Applying the ecological footprint to ecotourism scenarios. *Environmental Conservation,* 32, 294–304.

Hunter, C. and Shaw, J. (2007) The ecological footprint as a key indicator of sustainable tourism. *Tourism Management*, 28, 46–57.

Kerley, G.I.H., Geach, B.G.S. and Vial, C. (2003) Jumbos or bust: do tourists' perceptions lead to an under-appreciation of biodiversity? *South African Journal of Wildlife Research*, 33, 13–21.

Kimmel, J.R. (1999) Ecotourism as environmental learning. *Journal of Environmental Education*, 30, 40–44.

Kruger, O. (2005) The role of ecotourism in conservation: panacea or Pandora's box? *Biodiversity and Conservation*, 14, 579–600.

Kuo, I.-L. (2002) The effectiveness of environmental interpretation at resource-sensitive tourism destinations. *International Journal of Tourism Research*, 4, 87–101.

Lee, T. and Balchin, N. (1995) Learning and attitude change at British Nuclear Fuel's Sellafield visitors centre. *Journal of Environmental Psychology*, 15, 283–298.

Lee, W.H. and Moscardo, G. (2005) Understanding the impact of ecotourism resort experiences on tourists' environmental attitudes and behavioural intentions. *Journal of Sustainable Tourism*, 13, 546–565.

Lück, M. (2003) Education on marine mammal tours as agent for conservation – but do tourists want to be educated? *Ocean and Coastal Management*, 46, 943–956.

Madin, E.M.P. and Fenton, D.M. (2004) Environmental interpretation in the Great Barrier Reef Marine Park: an assessment of programme effectiveness. *Journal of Sustainable Tourism*, 12, 121–137.

Manfredo, M. (ed.) (1992) *Influencing Human Behaviour: Theory and Application in Recreation, Tourism and Natural Resource Management*. Sagamore, Champaign, IL.

Marion, J.L. and Reed, S.E. (2007) Minimizing visitor impacts to protected areas: the efficacy of low impact education programmes. *Journal of Sustainable Tourism*, 15, 5–27.

Mason, P. (2005) Visitor management in protected areas: from 'hard' to 'soft' approaches? *Current Issues in Tourism*, 8, 181–194.

Moscardo, G. (1996a) *Principles of Effective Interpretation*. CRC Reef Research, Townsville.

Moscardo, G. (1996b) Mindful visitors: heritage and tourism. *Annals of Tourism Research*, 23, 376–397.

Moscardo, G. (1999) *Making Visitors Mindful*. Sagamore, Champaign, IL.

Moscardo, G. (2003) Interpretation and sustainable tourism: Functions, examples and principles. *Journal of Tourism Studies*, 14, 112–123.

Moscardo, G. and Woods, B. (1998) Managing tourism in the Wet Tropics World Heritage Area: interpretation and the experience of visitors on Skyrail. In E. Laws, B. Faulkner and G. Moscardo (eds) *Embracing and Managing Change in Tourism: International Case Studies*. Routledge, London, pp. 307–323.

Mowforth, M. and Munt, M. (2003) *Tourism and Sustainability: Development and New Tourism in the Third World* (2nd edition). Routledge, London.

Niefer, I.A., da Silva, J.C.G.L. and Amend, M. (2002) Analysis of the visitors of Superagüi National Park, Brazil. *Current Issues in Tourism*, 5, 208–221.

Orams, M.B. (1994) Creating effective interpretation for managing interaction between tourists and wildlife. *Australian Journal of Environmental Education*, 10, 21–34.

Orams, M.B. (1995) Using interpretation to manage nature-based tourism. *Journal of Sustainable Tourism*, 4, 81–94.

Orams, M.B. (1997) The effectiveness of environmental education: Can we turn tourists into 'greenies'? *Progress in Tourism and Hospitality Research*, 3, 295–306.

Pearce, P.L. and Moscardo, G. (1998) The role of interpretation in influencing visitor satisfaction: a rainforest case study. In W. Faulkner, C. Tidswell and D. Weaver (eds) *Progress in Tourism and Hospitality Research* (Part 1). Proceedings of the Eighth Australian Tourism and Hospitality Research Conference. Bureau of Tourism Research, Gold Coast, Canberra, pp. 309–319.

Ritchie, B.W., Carr, N. and Cooper, N. (2003) *Managing Educational Tourism*. Channel View Publications, Clevedon.

Ryan, C., Hughes, K. and Chirgwin, S. (2000) The gaze, spectacle and ecotourism. *Annals of Tourism Research*, 27, 148–163.

Saleh, F. and Karwacki, J. (1996) Revisiting the ecotourist: the case of Grasslands National Park. *Journal of Sustainable Tourism*, 4, 61–80.

Sparks, P. and Guthrie, C. (1998) Self-identity and the theory of planned behaviour: a useful addition or an unhelpful artifice? *Journal of Applied Social Psychology*, 28, 1393–1417.

Staiff, R., Bushell, R. and Kennedy, P. (2002) Interpretation in National Parks: some critical questions. *Journal of Sustainable Tourism*, 10, 97–113.

Stein, T.V. and Lee, M.E. (1995) Managing recreation resources for positive outcomes: an application of benefits-based management. *Journal of Park and Recreation Administration*, 13, 52–70.

Stein, T.V., Denny, C.B. and Pennisi, L.A. (2003) Using visitors' motivations to provide learning opportunities at water-based recreation areas. *Journal of Sustainable Tourism*, 11, 404–425.

Tilden, F. (1977) *Interpreting our Heritage* (3rd edition). University of North Carolina Press, Chapel Hill, NC.

Tisdell, C. and Wilson, C. (2005) Perceived impacts of ecotourism on environmental learning and conservation: turtle watching as a case study. *Environment, Development and Sustainability*, 7, 291–302.

Tubb, K.N. (2003) An evaluation of the effectiveness of interpretation within Dartmoor National Park in reaching the goals of sustainable tourism development. *Journal of Sustainable Tourism*, 11, 476–498.

Wallace, G.N. and Pierce, S.M. (1996) An evaluation of ecotourism in Amazonas, Brazil. *Annals of Tourism Research*, 23, 843–873.

Wearing, S. and Neil, J. (1999) *Ecotourism: Impacts, Potential and Possibilities*. Butterworth Heinemann, Oxford and Melbourne.

Weaver, D.B. (2005) Comprehensive and minimalist dimensions of ecotourism. *Annals of Tourism Research*, 32, 439–455.

Weaver, D.B. and Lawton, L.J. (2007) Twenty years on: the state of contemporary ecotourism research. *Tourism Management*, 28, 1168–1179.

Wet Tropics Management Authority (1995) *Draft Wet Tropics Plan*. WTMA, Cairns.

White, C.J. (2005) Culture, emotions and behavioural intentions: implications for tourism research and practice. *Current Issues in Tourism*, 8, 510–531.

Wight, P.A. (1996) North American eco-tourism markets: Motivations, preferences and destinations. *Journal of Travel Research*, 25, 3–10.

Chapter 12
Broadening the View of Ecotourism: Botanic Gardens in Less Developed Countries

Lucy A. Sutherland

> We live in a world that is subject to continual change, often rapid, sometimes dramatic. These changes may be (a) political, (b) social, (c) environmental, (d) demographic, (e) climatic, (f) institutional. Not surprisingly, all kinds of organizations and institutions are learning to adapt so as to be able to live with these changes, and botanic gardens are no exception (Heywood 1996: 1).

Introduction

Over the past three decades, the push for sustainable forms of development have influenced and guided the work of contemporary botanic gardens. Around the world, botanic gardens are being reinvented and evolving into important centres for sustainability, in particular in relation to education and the maintenance and use of biodiversity (Wyse Jackson and Sutherland 2000). Tourism has become a particular focus for many botanic gardens since, in the late 1990s, the Secretariat of the Convention on Biological Diversity (CBD) identified this industry as a cross-cutting issue for the conservation and sustainable use of biodiversity.

Various tourism markets rely on the quality of a country's biodiversity. However, the industry, especially nature-based tourism, can place enormous pressure on it by expanding into remote and sensitive areas (Secretariat of the CBD 2004). Land managers, including botanic garden managers, are concerned about tourism pressure on protected and unprotected natural area systems, and subsequent impacts from the unsustainable use of biodiversity (Obua 1997, Buckley 1999, 2002, Sutherland 2000). Consequently, some botanic garden managers are examining the role that their institutions can play in ecotourism to support the protection of local biodiversity, build capacity of the community for the sustainable use of biodiversity, raise awareness about biodiversity values, as well as generate much needed revenue to support the maintenance of their botanic garden and the institution's work in conservation and sustainable development (Wyse Jackson and Sutherland 2000).

The idea of botanic gardens providing an ecotourism experience challenges the common principles in ecotourism paradigms. While there is no consensus on the use of the term, Fennell's (2001) content analysis of definitions highlights issues for botanic gardens in that ecotourism is defined in terms of occurring in natural areas, parks/protected areas or undeveloped areas for around half the definitions examined (n=85). In addition, the ecotourism experience is often portrayed as remote and lacking in crowds, combined with physically challenging recreational opportunities (Wight 1997); all of which are commonly associated with natural and protected areas, but not with botanic gardens.

As outlined in its introduction, this publication offers a conceptual and practical insight into the complex interactions between ecotourism and heterogeneous and dynamic natural environments. This chapter challenges paradigms that present a limited ecotourism view in terms of pristine natural environments. It is argued that the ecotourism experience is both personally and socially constructed and, consequently, there is potential for botanic gardens in Less Developed Countries (LDCs) to be viable settings for some ecotourists. The cases presented illustrate that some botanic gardens in LDCs already have a role in ecotourism, and further development of their role could alleviate pressure on protected areas by providing an alternative ecotourism experience. Moreover, the botanic garden experience can complement the protected area experience through education and awareness-raising programmes that prepare tourists for their protected area experience, as well as address appropriate tourism behaviour in nature settings. The chapter concludes that, at a broader level, the socio-politics of the host country is a key factor hindering the integration of botanic gardens into ecotourism; and, at a grass roots level, the passive approach taken by their management allows the tourism industry's western view of botanic gardens in LDCs to persist, with the consequence that this social construction of botanic gardens limits their recognition as viable settings for ecotourism.

Ecotourism and Modified Spaces

Several seminal ecotourism works centre on researching or presenting ecotourism in the context of national parks/protected areas (for example, Boo 1990a and b, Brandon 1996, Ceballos-Lascuráin 1996, Honey 1999). However, ecotourism debates in the literature have struggled when considering the 'margins' and opinions on activities and operations that do not clearly fit into mainstream ecotourism thinking (Orams 2001). Nevertheless, there is growing academic discourse on ecotourism and modified spaces that attempts to broaden current views by exploring ecotourism as an adaptive paradigm in terms of different settings (Chirgwin and Hughes 1997, Mason 2000, Lawton and Weaver 2001, Weaver 2001, Higham and Lück 2002, Ryan and Saward 2004, Chirgwin 2005, Weaver 2005).

Several scholars argue in favour of leaving pristine areas free from ecotourism and using areas, such as modified spaces, that have a higher carrying capacity for this market (Mason 2000, Higham and Lück 2002, Weaver 2005). Mason (2000), for example, claims that zoos, commonly located close to major tourist regions, are more appropriate for ecotourism and provide tourism that is specialized, educational and nature-based, without involving the heavy consumption of finite resources through long-distance travel. Higham and Lück (2002) also present the case for ecotourism within an urban context and highlight the difference between academic discourse that focuses on ecotourism in pristine areas and the practical reality of commercial ecotourism operations, which often use modified spaces for wildlife-focused ecotourism. Furthermore, these authors examine the contradictions and constraints posed by ecotourism paradigms that focus on unmodified, natural and pristine settings, and question whether measures of naturalness should be considered in ecotourism.

Along similar lines, Chirgwin (2005) argues that 'nature-based' is a more appropriate criterion than 'natural', as it allows for heavily modified and rehabilitated sites to be accepted for ecotourism because of the wildlife viewing opportunities they provide. Her case study research revealed that modified spaces were viable settings for nature-based tourism and satisfied the nature tourist's motivations and expectations. Waitt et al. (2003) also examined the boundaries of nature tourism and found that tourists perceived modified settings, such as dams, as part of the natural landscape. These works highlight that ideas of nature are personally constructed concepts and people's perceptions of what is 'natural' evolve and are determined by their previous experiences (Waitt et al. 2003).

Ecotourism has been discussed in terms of other modified settings, including artificial reefs, agricultural lands and surface mines and quarries (Lawton and Weaver 2001, Weaver 2001). While all of these works make an important contribution to the discussion on ecotourism and modified spaces, several are limited because they have tended to construct ecotourism with a wildlife focus. With the exception of Weaver (2005), an ecotourism experience in terms of plants is rarely discussed. Furthermore, botanic gardens have been discounted from having a role in ecotourism because of their 'captive nature' (Lawton and Weaver 2001, Weaver 2005). This is arguably a western constructed view of botanic gardens, where early European botanic gardens were based on a formal landscape design and a philosophy 'of power over nature' (Prest 1981: 54). This contrasts with contemporary botanic gardens, where around 400 of the world's 2000 botanic gardens include natural vegetation and/or natural areas (Wyse Jackson and Sutherland 2000).

Botanic gardens in LDCs are often located in regions with rich biodiversity that are experiencing a growth in nature-based tourism. With increasing discussions on types of ecotourism, for example deep and shallow (Acott et al. 1998); hard and soft (Laarman and Durst 1987, cited in Orams 2001: 28); and comprehensive and minimalist (Weaver 2005), as well as the growing consideration of modified spaces for ecotourism, there is a need within this discourse to consider the role of botanic gardens in LDCs.

Institutions Evolving for Change: A Context for Botanic Gardens and Ecotourism

Botanic gardens in LDCs are not homogeneous attractions and their potential for ecotourism is consequently variable. However, in their favour, botanic gardens are seen as capable of adaptation (Bolanos 1987, Kohlmaier and von Sartory 1990, Wallace 1997, Willison 1997, Oikawa 2000). Oikawa (2000) claims that by evolving and adapting to world circumstances and societal needs, the existence of botanic gardens can be justified in contemporary society. While early botanic garden development in LDCs was intrinsically linked with colonization and arguably used as a tool for globalization (see Brockway 2002), more recently their work has focused on conservation and sustainable development.

Ecotourism is often discussed in terms of conservation (for example, Honey 1999, Fennell 2001), and the evolving work of these botanical institutions complements this conservation ideology within ecotourism. There are numerous cases where botanic garden development has involved significant habitat rehabilitation and, in addition, many link into important urban green space and wildlife corridor networks. Furthermore, their work in *ex situ* and *in situ* conservation at a local, regional and international level is increasing (Guthrie 1996, Dixon 1997, Maunder and Culham 1997).

Benefits to local people are also seen as a key component of around half of the ecotourism definitions examined by Fennell (2001) and, in terms of their operations, many botanic gardens contribute to local sustainable development. A preliminary survey in LDCs, by the author, revealed botanic gardens are important employers of local people and for more than one-fifth (22 per cent, n=110)[1] all of their staff were born in the local area (that is within a 20 km radius) and for two thirds (66.4 per cent, n=110) of botanic gardens, half or more of the garden staff lived locally. In addition, more than half (59.1 per cent, n=115) of the respondents from LDC botanic gardens indicated that local businesses and industry benefit from the presence of the botanic gardens through an increase in tourism to the area. Furthermore, botanic garden managers regularly use services provided by local business and industry, for example machinery hire, trades persons, printing, promotion and catering.

In terms of communities, strong partnerships with botanic gardens in LDCs are well documented (for example, Pushpangadan et al. 1998, Hariramamurthi 2000, Symonds 2000, 2001, Ewane Sumelong 2001, Seeni et al. 2002). The preliminary survey of LDC botanic gardens by the author revealed that more than two thirds (68.9 per cent, n=122) are engaged in community programmes. Some of this work contributes to the broader goal of sustainable development and involves community partnerships, and education and capacity-building of youth and minority groups.

1 There were 123 responses to this survey. However, some questions were not completed by all respondents and therefore 'n' indicates the number of responses to the particular question being discussed.

In addition, most staff (89.1 per cent, n=119) believed that individual community members benefit from the botanic gardens' existence and, more specifically, they benefit from educational services and public learning programmes (53.5 per cent, n=101), recreational opportunities (38.6 per cent), the plant nursery (11.9 per cent) and the provision of a natural area in an urban environment (6.9 per cent).

From a practical perspective, ecotourism is sometimes defined as minimum impact tourism (Fennell 2001). While a detailed discussion on this topic is beyond the scope of this chapter, it should be noted that botanic garden operations are being revised to incorporate the principles of environmental sustainability including composting garden waste, water conservation programmes, reducing/eliminating chemical use, and using local and organic produce in their restaurants (Sutherland 2005).

From the above discussion, there is evidence that botanic garden operations complement the principles that ecotourism is commonly thought to uphold. However, ecotourism in botanic gardens also needs to be considered from the tourism industry perspective to investigate how these attractions, and the experiences they provide, are perceived in terms of ecotourism principles.

Gathering Evidence for Ecotourism and Botanic Gardens

This chapter draws on findings from multi-strategy research to explore botanic gardens for ecotourism in LDCs, and examines factors that challenge their integration into the industry. The research involved a study of Kirstenbosch National Botanical Garden (KNBG) in South Africa and Jardín Botánico Nacional de Cuba (JBNC) in Cuba. More specifically, tour operator and tourist perspectives were examined to investigate the potential role of these botanic gardens in the ecotourism sector (Table 12.1).

Based on discussions with botanic garden staff, a mail survey using Dillman's technique (1978) was administered to Cape Town tour operators and an email survey administered to Havana-based operations to obtain quantitative data on the use and opinions of the botanic gardens by tour operators. In addition, qualitative data was collected using semi-structured interviews with selected tour operators in both capital cities. Furthermore, an interview-administered tourist survey in the gardens sought information on tourist expectations of settings, perceptions of experiences and motivations for visiting botanic gardens. This survey was replicated in nearby protected areas (Table Mountain National Park (TMNP) in South Africa and Biosfera Sierra Del Rosario (SDR) in Cuba) to provide comparative national park data on tourist expectations, perceptions and motivations.

Table 12.1 Number of interviews conducted

	South Africa	Cuba
Tourists in botanic gardens	196	105
Tourists in protected areas	90	31
Botanic garden staff	3	2
National park staff	2	3
Tour operators	5	5
Representatives from local and regional tourism industry	2	0

Note: While the number of days spent surveying did not vary (17 days in botanic gardens and 9 days in protected areas), the number of respondents was determined by obvious factors such as the number of tourists visiting on the day, tour group demands and tourists' time availability. For example, tourists with children were hesitant to spend time being interviewed.

Background to Tourism in the Case Study Botanic Gardens

Before exploring the subject of ecotourism and botanic gardens, it is important to briefly introduce the tourism infrastructure and services in KNBG and JBNC.

Kirstenbosch National Botanical Garden (KNBG)

During the late 1990s, the visitor facilities at Kirstenbosch were upgraded to an international standard and the site is now managed using primarily a neoliberal model. KNBG staff and volunteer guides provide most information and interpretive services, while commercial operations are outsourced to independent operators and KNBG receives a percentage of their turnover. The 528 ha garden has a well-defined place in mainstream tourism and is one of South Africa's top ten attractions for overseas tourists (SA Tourism cited in Grant Thornton Kessel Feinstein (GTKF) 2001). Strong partnerships with the tourism authorities and other tourist attractions exist and KNBG is one of the founding members of the Big Six Forum, an informal network of primary tourist attractions in the Western Cape, which meets regularly to explore cross-marketing opportunities (NBI 2000). Consequently, by integrating into the tourism industry, KNBG has reduced its financial dependence on the state and has provided the South African National

Biodiversity Institute with a model of good practice for its other botanic gardens (NBI 2003).

Jardín Botánico Nacional de Cuba (JBNC)

The 600 ha JBNC has well-developed tourism and visitor infrastructure, including Havana's first vegetarian organic restaurant, as well as cafes, a gift shop, and car/coach parking. All its services are state managed and funded and a proportion (around one third) of any tourism profits are returned to the state treasury. A guiding service is available for all visitors on their arrival and JBNC employs 17 full-time guides, who are multi-lingual and university trained in education and biology. In contrast to KNBG, JBNC is not well integrated into the international tourism industry. As discussed later in this chapter, its role in international tourism is limited by Cuba's socio-politics and, consequently, JBNC's programmes focus on science and education for the benefit of Cuban nationals.

Examining the Potential for Botanic Gardens in Ecotourism

The Current Status

Visitation to KNBG and JBNC is dominated by the domestic market and international tourists contributed 23.2 per cent and 4.6 per cent, respectively, to the total botanic garden visits in 2002. Despite the smaller proportion of international tourist visits, both gardens attract international ecotourism groups, specifically for their avifauna and native flora. Their key roles in ecotourism focus on education, training and capacity building, as well as providing a setting for introductory learning experiences (Table 12.2).

To examine the further potential of these botanic gardens in LDCs for ecotourism, they will be considered in terms of the components of an ecotourism experience, the quality of the nature experience and the learning opportunities that they provide.

Perceptions of the Botanic Garden Experience

Initially, the botanic gardens will be examined in terms of their effectiveness in meeting some of the key experiences commonly associated with ecotourism (Wight 1997), for example:

1. Remoteness from civilization.
2. An uncrowded experience.
3. An opportunity to learn about plants/wildlife/nature.
4. An opportunity to view animals and plants.
5. Recreational opportunities that offer a physical challenge.

The potential of botanic gardens to provide these experiential elements was highlighted during surveys. Figure 12.1 illustrates that, for some tourists, botanic gardens effectively provide an experience that is remote and lacks crowds, with nature-focused learning and plant viewing opportunities. Approximately half of the respondents at KNBG and JBNC rated a botanic garden experience generally as remote and over 55 per cent rated it as uncrowded. Furthermore, the nature focus in botanic gardens was evident, with greater than 90 per cent of the respondents rating these attractions as effective in providing opportunities to view plants. A further 80 per cent (KNBG) and 90 per cent (JBNC) of respondents rated botanic gardens as effective in providing opportunities to learn about plants/wildlife and nature.

Key weaknesses in the botanic garden experience, in terms of ecotourism, are the perceived limited opportunities to view wildlife and to undertake physical challenges (Figure 12.1). Few KNBG (21 per cent) and JBNC (37 per cent) respondents rated botanic gardens as effective in providing opportunities to view wildlife (including birds) and only around one third rated them as effective in

Table 12.2 Involvement of case study botanic gardens in the ecotourism industry

Kirstenbosch National Botanic Garden (KNBG) South Africa	Jardín Botánico Nacional de Cuba (JBNC) Cuba
• Small-scale, nature-focused South African operators use KNBG for an introductory session to birds and South African flora before visiting protected areas.	• Havanatur, a Cuban state owned tour company, uses JBNC and its staff to introduce tourists to Cuba's biodiversity (including avifauna) before the protected area(s) visit.
• Provides specialist garden guides.	• Provides specialist multi-lingual garden guides.
• Staff conduct biodiversity training, when requested, on flora and conservation for accredited tour guides.	• Staff conduct training courses, when requested, on flora and conservation for Cuban accredited tour guides.
• Regular feature articles in the tour guide newsletter.	• JBNC has themed trails in its natural areas for the sole use of ecotourism groups.
• Exposure of ecotourism operators via KNBG's Peninsula Tourism Kiosk.	

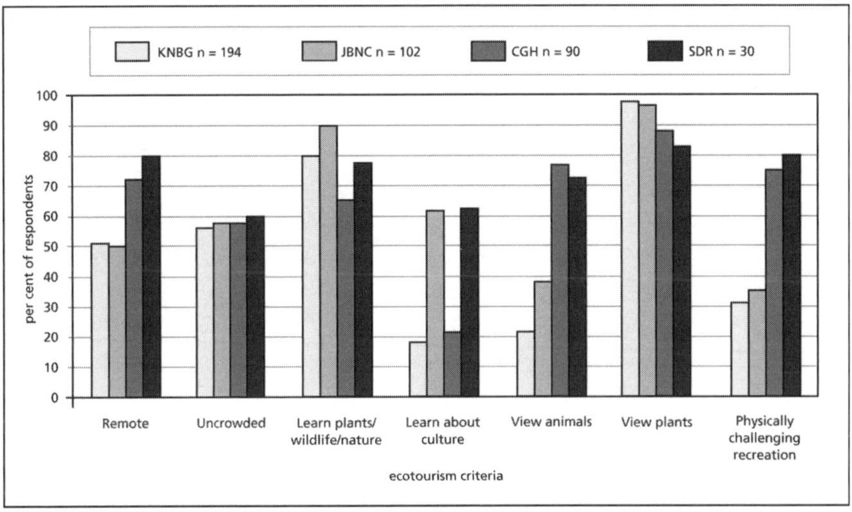

Figure 12.1 Respondents ranked botanic gardens or national parks generally as effective in providing selected components of an ecotourism experience

providing physically challenging recreation. However, this is a perception of botanic gardens *generally* held by sampled tourists, and the reality in individual gardens can be vastly different. Both KNBG and JBNC are important wildlife corridors and sites for bird watching. Furthermore, KNBG is part of the Cape Floristic Region Protected Area and based at the foot of Table Mountain. KNBG tourists can take a strenuous hike from the plant collection to the mountaintop through natural ravines (Figure 12.2). There are clearly weaknesses, therefore, in the argument that botanic gardens are captive nature and inappropriate for ecotourism.

The research also revealed that the sampled tourists perceived national parks generally as ineffective in providing all the core elements of an ecotourism experience (Figure 12.1). Interestingly, more than 40 per cent of respondents did not rank national parks as effective in providing an uncrowded experience. In addition, they were not rated as effective as botanic gardens in providing opportunities to learn about plants/wildlife and nature. Nevertheless, the strengths of national parks for ecotourism are demonstrated by the findings that around 70 per cent to 80 per cent of respondents perceived them as effective in providing a feeling of remoteness from civilization, offering opportunities to view animals and offering physically challenging recreation.

Figure 12.2 Natural areas within Kirstenbosch, and the adjacent Table Mountain National Park, can provide physically challenging recreational opportunities for tourists

Source: Gigi Laidler, South African National Biodiversity Institute.

These findings highlight some of the limitations of stereotyping settings for ecotourism. From this study of tourist perceptions, there is evidently scope for some botanic gardens to play a greater role in ecotourism, recognizing that diverse life experiences contribute to people's perceptions of experiences and settings. These findings raise questions as to the role of botanic gardens in forms of ecotourism such as 'hard' and 'soft', and 'deep' and 'shallow'. The next section examines botanic gardens more specifically in terms of their nature experience.

Quality of the Botanic Garden Nature Experience

Several authors have contributed to discourse on the concepts of 'natural' and 'nature' and how humans construct these (Tuan 1990, Macnaghten and Urry 1998, Higham and Lück 2002, Chirgwin 2005). The concept of 'natural' is often linked to ecotourism (see Fennell 2001) and, therefore, botanic garden settings need to be considered in terms of their natural/nature experience. As the literature highlights, 'nature' and 'natural' are perceived in many ways and this was emphasized when approximately one quarter (28.3 per cent, n=124) of tourists described KNBG's setting as 'natural'. More commonly, around half (49.2 per cent, n=124) described it as a semi-natural/semi-cultivated setting, and for others (22.5 per cent, n=124), it was a cultivated landscape. In contrast, a greater proportion of tourists (62 per cent, n=79) described JBNC as a 'cultivated' setting, whilst few (5.1 per cent) perceived it as 'natural'. The different perceptions of these rehabilitated pastoral landscapes may be associated with the tourists' previous nature experiences, the age of botanic garden development (where KNBG was established in 1913 and JBNC in the late 1960s), or with the site characteristics (for example, JBNC is situated on relatively flat terrain and KNBG has Table Mountain as a dramatic backdrop) and how these are interpreted by the tourist (Figure 12.2).

While only a small proportion of tourists described either garden setting as 'natural', the overlap of the concepts of 'natural' and 'nature' provide scope for botanic gardens in ecotourism in terms of their nature experience. In South Africa, the great majority of respondents (92.6 per cent, n=122) and South African tour operators (93.8 per cent, n=32) ranked the quality of KNBG's nature experience as excellent. In addition, more than two-thirds (69.7 per cent) ranked it as an 'important' or 'very important' nature tourism destination. Similarly, the majority (79.7 per cent, n=79) of JBNC respondents ranked the quality of its nature experience as excellent, as did tour operators (n=3) during interviews. Two Cuban operators also indicated that JBNC was an 'important' nature-based destination, in comparison to other destinations that they offered.

These findings support the argument for the integration of botanic gardens in LDCs into ecotourism. The final aspect of botanic gardens to be considered in terms of ecotourism is education, because it is commonly associated with ecotourism and it is included in various definitions (41.2 per cent, n=85) (Fennell 2001).

Education and Learning Opportunities

The role of KNBG and JBNC in ecotourism mainly involves providing a nature-based setting and the learning/awareness raising/interpretation element of ecotourism. Within ecotourism paradigms, education and learning has been viewed as part of the tourism experience in a passive way to enhance satisfaction and enjoyment, or in an active way to contribute to protecting resources by modifying tourist behaviour and attitudes (Ritchie et al. 2003). The tourist education activities at KNBG and JBNC are primarily modelled on ecotourism paradigms with a passive approach, and are hence non-outcome based programmes. For the most part, tourism satisfaction and enjoyment is their focus. This passive educational approach for tourists contrasts with the active approach taken with programmes for nationals (Hernández 1992, Ashwell and Bobo-Mrubata 1998, Betancourt et al. 2000, Ashwell 2001). Therefore, within these institutions, there are skills and knowledge available to support a proactive approach to tourism education.

Consequently, there is potential for botanic gardens to expand their educational/interpretive role in ecotourism. At KNBG and JBNC, for example, the garden visits are often positioned at the beginning of an ecotour itinerary, prior to the group visiting protected areas (Table 12.2). This visit can thereby educate tourists about the biodiversity that they will see in more 'natural' habitats, support the protection of local biodiversity through raising awareness of its values, and educate tourists on appropriate ways to enjoy natural environments whilst still protecting biodiversity.

Hindrances to Botanic Gardens in Ecotourism

The previous section examined the potential of botanic gardens in LDCs for ecotourism in terms of the experiential components of this type of tourism, the botanic garden nature experience, and the educational and learning opportunities provided by these attractions. The discussion indicates that botanic gardens such as KNBG and JBNC are well placed to expand into the ecotourism industry, particularly considering that for some tourists these settings effectively provide a feeling of remoteness, which lack crowds and provide learning opportunities. In addition, for a great majority, they are excellent nature experiences. However, the research also revealed that the ability of botanic gardens to integrate into the ecotourism industry is hindered by several factors. The following discussion focuses on three such factors and discusses them with particular reference to the case study research in South Africa and Cuba:

1. The socio-political situation of the host country
2. The management approach of the botanic garden to developing and monitoring their role in ecotourism
3. Industry perceptions of botanic gardens.

Host Country Socio-politics

Brockway (2002) highlights the relationship between botanic gardens and politics in her historical reflection on the science role of British Royal Botanic Gardens in colonial exploration. Today, politics continues to determine the development and management of most botanic gardens, and Michel (1996) and Morris (1997) illustrate how political circumstances have led to the decay of several botanic gardens in LDCs. In terms of tourism, politics influences the integration of botanic gardens into the broad industry and more specifically ecotourism. For example, politics has resulted in the Jardín Botánico Nacional de Cuba being pigeonholed into a traditional botanic garden role in science and formal education. Since opening to the public in 1984, the mission of JBNC reflects Cuba's priority to conduct programmes that benefit nationals:

> … to promote knowledge and awareness of the plant life among population, emphasising Cuban native flora, contributing to its conservation and

> investigation, as well as to teaching botany at different educational levels (Leiva Sánchez 1996: 61).

As a consequence, JBNC's programmes in sustainable development and conservation, and its well-established visitor facilities and services, primarily cater for Cuban nationals. The desire of JBNC management to expand its focus and engage with international tourists links, firstly, with the global growing interest of botanic gardens in the implementation of the Convention on Biological Diversity and, secondly, with Cuba's 1993/1994 reform to engage with international tourism in response to losing economic support following the collapse of the Soviet Union. International tourism is a revenue-raising tool for JBNC to support its work in conservation and sustainable development.

However, the role of JBNC in international tourism is limited because it is under the Universidad de la Habana management structure and is thereby constrained by its science and education role. Attractions under the Ministry of Tourism benefit from marketing and promotion campaigns, as well as being profiled as Havana's icon tourist attractions. In contrast, a content analysis of Cuban tourism publications collected from the 2002, 2004 and 2006 World Travel Markets in London revealed little attention paid to JBNC, or other attractions that were not within the Ministry's portfolio. Under Cuba's political regime, JBNC staff are prohibited from liaising directly with international guide publishers and tour operators and, consequently, this challenges their attempts to promote the garden to an international market and engage with the ecotourism industry.

KNBG management, by contrast, faces different challenges. While the focus of KNBG's work over nearly a century shows an inward-looking botanical institution (see South African National Biodiversity Institute 2008), post-apartheid saw the government's Reconstruction and Development Programme, including strengthening its tourism industry, form the basis of the institution's activities. Politics has enabled management to expand KNBG's role in tourism and to integrate it into many Cape Town tourism products while, at the same time, its high profile challenges its ability to have a role in ecotourism.

In summary, both JBNC and KNBG battle for a position in the ecotourism industry because of the politics of their host countries. JBNC struggles because the political structure has it positioned as a scientific and educational institution and thereby restricts the ability of its management to integrate it into ecotourism. The high profile of KNBG in the mainstream tourism industry, by contrast, creates strong images of this garden that contrast with perceptions of an ecotourism experience and, therefore, limit its ability to integrate into this industry.

The Approach of Management to Developing and Monitoring the Role of JBNC and KNBG in Ecotourism

While, arguably, host country socio-politics can be a significant hindrance to botanic gardens engaging in ecotourism, as discussed above, the management of botanic gardens also needs to be acknowledged in limiting integration into the industry. Despite the interest of their management in ecotourism, the two case studies tend to operate on the periphery of this industry, with no evidence of their staff being key players or proactive in ecotourism. Areas needing greater attention are the strategic planning commitment to ecotourism, and being proactive in biodiversity and botanic garden-focused education programmes for industry representatives and ecotourists.

There have been some strategic attempts to develop the roles of KNBG and JBNC in ecotourism, although these are still at the early stages of development. KNBG's strategic planning shows a fundamental shift in focus towards nature-based tourism, compared to earlier strategies. One of the institute's goals is being 'A key player in the nature-based tourism industry' and a core business development area is developing 'ecotourism destinations' (NBI 2003: 6). This strategic vision, however, has not yet been put into practice and KNBG tends to react to demand from nature-based and ecotourism operators, instead of taking the proactive approach it takes with the mainstream tourism industry. In contrast, JBNC's Marketing and Promotion Officer has developed one-day tour packages for niche markets including ecotourists. The promotion of these, however, is limited to delivering information packs and conducting meetings with Havana based operators; direct liaison with overseas operators is prohibited. Consequently, the Jardín Botánico Nacional is described as 'one of Havana's many well-kept secrets' (The Economist (US) 2000: 87).

As previously mentioned, and from an operational perspective, both KNBG and JBNC need to be proactive in the provision of tourist learning programmes. KNBG's management body, the South African National Biodiversity Institute (SANBI), for example, aims 'to inspire and empower people [nationals] to take responsibility for their environment' through its environmental education programmes (NBI 2003: 36), yet no such educational aims are set for tourist programmes. Instead, the focus is on creating a 'good public image … increasing awareness of [the] gardens and encouraging visits to them' (NBI 2003: 40). According to Redclift and Sage (1994) the way people value nature influences their ability to manage the environment for sustainability, and so education in ecotourism needs to form part of a planned programme of learning for it to contribute to the development of tourist values and environmentally friendly behaviour appropriate for ecotourism.

JBNC and KNBG staff also have a role in the training and capacity building of nature-based tour guides (Table 12.2). However, again the provision of this tourism service was found to be reactive to demand. Taking a greater role in training and capacity building of nature-based tourism guides would raise the profile of these gardens for ecotourism through greater understanding of the opportunities and experiences that they can provide.

Industry View of Botanic Gardens

The perception of botanic gardens held by the tourism industry also has a significant influence on how these attractions are used, marketed and promoted for ecotourism. The passive approach taken to ecotourism by botanic garden management arguably results in many tour operators and industry representatives exhibiting a narrow view of botanic gardens and their potential for ecotourism. The view of the tourism industry as a hindrance to botanic gardens for ecotourism is discussed now in terms of the botanic garden image and stereotype and the nature experience provided.

Weak image

While botanic gardens are biodiversity rich, as one South African tour operator explained, their image is not strong enough to sell a country or even a destination. This contrasts with national parks, such as Kruger, Uluru-Kata Tjuta, and Serengeti, which can do both. This weak image of botanic gardens was emphasized in interviews with tour operators and, in addition, it was reiterated during tourist interviews when few KNBG tourists (KNBG 8.9 per cent, n=190) indicated that the presence of botanic gardens was important in influencing their decision to select a holiday/vacation destination. In contrast, a significantly greater proportion of protected area respondents in South Africa ranked national parks higher in influencing their holiday destination selection (24.4 per cent CGH, n=90). ($\chi2 = 11.623$, df = 1, $p < 0.001$). Similarly, the Cuban study revealed that national parks were significantly more important in influencing tourists' holiday destination selection when compared to botanic gardens ($\chi2 = 6.701$, df=1, $p = 0.01$).

The weak image of botanic gardens has resulted in small nature based tour operators in South Africa encountering reluctance in their clients to visit botanic gardens:

> SA Operator 1: We haven't had people that really want to go to KNBG and really make an in-depth … or more than just a superficial acquaintance with it.

> SA Operator 2: … people are surprised [when they visit] because they [initially] think '… oh how boring we are going to a botanic garden', they get the huff you know?

This reluctance to visit predominantly stems from children and young adults, because botanic gardens are not perceived as providing active recreational or wildlife viewing opportunities:

> Male tourist: … they [teenagers] would go if they [botanic garden] offered more activities, like when we went to Costa Rica they did paint balling in the forest. Botanic gardens are mainly about 'looking' and for teenagers they need more activity.

> SA Operator 3: I've got the feeling that if you don't force them [younger people] a bit to go there, they won't go by themselves ...

Furthermore, botanic gardens are perceived as something to be 'grown' into and, therefore, associated with an older age group:

> SA Operator 3: People love to be in the garden … in France the older you are, the more you are interested [in] plants.

> SA Operator 4: … it is more [for] 50, 55 even 60 or 70 [age group]. I suppose because people that age are probably retired they've got more time to have a passion and spend time in the garden.

Tour operators also perceived limitations in botanic gardens. Operators considered the appeal of KNBG and JBNC for ecotourism negligible. Of the South African operators surveyed (n=27), only one saw KNBG appealing to ecotourists. Similarly, only one out of three Cuban operators interviewed considered JBNC had a role in ecotourism.

Nevertheless, this chapter presents data to support the argument that the concept of ecotourism and modified spaces needs to be broadened and, if one quarter of tour operators consider KNBG had appeal to the scientific/botanical (25.9 per cent) and/or nature (22.2 per cent) niche markets, then surely there is potential for ecotourism?

Western stereotype of settings

Associated with the image of botanic gardens, is the setting that they provide. Content analysis undertaken by Fennell (2001) revealed that the greater majority (91.8 per cent) of ecotourism definitions did not refer to the scale of development. Yet there is a stereotype of ecotourism taking place in remote, pristine places, which implies that the scale of development is minimal and the setting is natural. Consequently, expectations of the setting and design of botanic gardens constitutes a worthwhile consideration when examining them in the context of ecotourism. Byrd (1989) explains that early European botanic gardens in the 1500s and 1600s were based on Renaissance thinking where the natural world was to be controlled, and their designs reflected the development of mathematics as a science. By the mid-1700s, further order of plant collections was directed by Linnaeus's binomial nomenclature system (Byrd 1989). While the layout of botanic gardens has evolved from these early design models to plantings that often reflect ecosystems and local habitats, the perception of botanic gardens as formal settings prevails because many European botanic gardens are still based on these early traditional designs.

Despite many of the world's botanic gardens having natural areas of vegetation (Wyse Jackson and Sutherland 2000), most respondents had strong views of botanic gardens as heavily modified settings. Discussions with tourists on the expectations of botanic garden and national park settings produced some predictable differences. National park settings are expected to be scenic and have rugged topography, wildlife and little development. In contrast, botanic gardens are expected to have settings associated with urban environments and include plant labels, displays of flowering plants, formal garden beds and water features (Sutherland 2005). Interestingly, during the field research tourists commented that the case studies were an exception to what they expected and broke the traditional botanic garden mould in terms of size, formality and naturalness, as the following tourist responses to JBNC illustrate:

> Denmark: I didn't expect it … our garden in Copenhagen is small ... the size of a postage stamp!
>
> Germany: I didn't expect the national botanic garden to be so big and so natural.
>
> Belgium: This is the first one that I've been to that is so big. I expect them to be very small.
>
> USA: I expected the botanic gardens to have more flowers and to be more formal.
>
> USA: I expected the botanic garden to be more formal, like the US gardens.

This strong reaction illustrates that some botanic gardens in LDCs do contrast with the traditional stereotype and it raises the question as to why these modified spaces have been overlooked in ecotourism discourse.

Concluding Argument for Ecotourism and Botanic Gardens

This chapter has presented evidence to support the argument that botanic gardens in LDCs have the potential for ecotourism. As argued by previous authors on modified spaces and ecotourism, urban ecotourism provides the possibility of alleviating pressure on fragile natural environments. Moreover, botanic gardens in LDCs challenge the traditional stereotype of botanic gardens as small, formally designed spaces; instead, many present large expanses with pockets of rehabilitated and natural vegetation. Furthermore, the botanic garden experience is associated with a high quality nature experience for many tourists and operators. In addition, the commitment to sustainable development and education provides a basis on which to build, as they integrate into the ecotourism industry.

Consider the position of American national parks 40 years ago, when they were not managed for a range of visitor experiences until the Recreational Opportunity Spectrum was devised (see Clarke and Stankey 1979, Driver et al. 1987). Initially there was little attention given to understanding people (Graham

1992), until extensive research revealed that people's perceptions and values of nature varied (for example Schreyer and Roggenbuck 1978, Schreyer and Beaulieu 1986, Tuan 1990, Manfredo et al. 1996). Moreover, it was largely a controlled view of nature that was presented. Are botanic gardens in any different a position than the national parks of the past? Can more credence be given to the individual to draw on their experience and knowledge and decide if botanic gardens provide an ecotourism experience? It must be acknowledged that not all national parks are appropriate for ecotourism, and similarly the potential ecotourism role of botanic gardens depends on such factors as their natural/nature setting, operational scale, level of development and location. Of equal importance, their potential depends on the tourists, their life experiences and the type of ecotourism being sought and promoted. All of these factors must be considered before claiming that botanic gardens have limited opportunities for ecotourism.

In conclusion, the words of British sculptor Richard Deacon are used to reflect on the reality of the national park ecotourism experience:

> We can never 'see' an object in its entirety, so in a sense only a fragment of what we see is the 'real' experience and part is remembered … and part is an 'imagining' of the form. In between, we have to make a visual and conceptual leap of faith to interpret the spatial complexities of a form (Richard Deacon cited in the National Gallery in London).[2]

During the ecotourism experience, we never see the ecosystem in its entirety, instead we see a fragment chosen by our ecotour operator and interpreted by our guide, who determines the story/ies we are told. It is then up to us to complete the image of the landscape to form our 'real' nature experience. So why then, if we do not experience and see the ecosystem in its entirety, should botanic garden settings be rejected from giving an ecotourism experience that is similarly socially and personally constructed? This chapter has shown that there are just as many reasons to utilize botanic gardens for ecotourism as there are to reject them for this form of tourism activity.

References

Acott, T., La Trobe, H.L. and Howard, S.H. (1998) An evaluation of deep ecotourism and shallow ecotourism. *Journal of Sustainable Tourism*, 6, 238–253.

Ashwell, A. (2001) Part 2: The NBI's vision and response. In J. Mkefe and A. Ashwell 'Botanical gardens education in the new South Africa: towards just provision of educational opportunity'. In C. Hobson and J. Willison (eds) *Teaching for the 21st Century: Botanic Gardens Education for the New Millennium*. Proceedings of the Third International Congress on Education in

2 Used with special permission of the artist given on 10 March 2008.

Botanic Gardens, Brooklyn Botanic Garden, New York City, 7–11 September 1996. Botanic Gardens Conservation International, USA, pp. 33–36.

Ashwell, A. and Bobo-Mrubata, P. (1998) Environmental action for change. Roots 16 *Botanic Gardens Conservation International Education Review*, July 1998, 24–26.

Betancourt, J., Acosta, D., Granados, A., Durán, M.B. and Durán, M.J.B. (2000) Proyecto educativo comunitario: comunidad alerta. *Revista del Jardín Botánico Nacional*, 21, 109–114.

Bolanos, C.A. (1987) Preface. In D. Bramwell, O. Hamann, V. Heywood and H. Synge (eds) *Botanic Gardens and the World Conservation Strategy*. Academic Press, London, pp. xv–xvi.

Boo, E. (1990a) *Ecotourism: The Potentials and Pitfalls* (Vol. 1). World Wildlife Fund, Washington DC.

Boo, E. (1990b) *Ecotourism: The Potentials and Pitfalls* (Vol. 2). World Wildlife Fund, Washington DC.

Brandon, K. (1996) Ecotourism and conservation: a review of key issues. *Environmental Department Papers, Biodiversity Series*, No. 033, Ecologically Sustainable Development. The World Bank, USA.

Brockway, L. (2002) *Science and Colonial Exploration: The Role of the British Royal Botanic Gardens*. Yale University Press, New Haven, CT and London.

Buckley, R. (1999) Tourism and biodiversity: land-use, planning and impact assessment. *Journal of Tourism Studies*, 10, 47–56.

Buckley, R. (2002) Tourism and biodiversity in North and South. *Tourism Recreation Research*, 27, 43–51.

Byrd, W.T. (1989) Re-creation to recreation. *Landscape Architecture*, January 79, 42–51.

Ceballos-Lascuráin, H. (1996) *Tourism, Ecotourism and Protected Areas: The State of Nature-Based Tourism around the World and Guidelines for its Development*. IUCN, Gland Switzerland and Cambridge UK.

Chirgwin, S. (2005) Can sites formerly subjected to development provide satisfying nature tourism experiences? Two case studies from the Top End of Australia's Northern Territory. *Journal of Sustainable Tourism*, 13, 50–62.

Chirgwin, S. and Hughes, K. (1997) Ecotourism: the participants' perceptions. *Journal of Tourism Studies*, 8, 2–7.

Clarke, R. and Stankey, G. (1979) *The Recreational Opportunity Spectrum: A Framework for Planning, Management and Research*. Pacific Northwest Forest and Range Experiment Station, General Technical Report PNW–98. Department of Agriculture and Forest Service, US.

Dillman, D. (1978) *Mail and Telephone Surveys*. John Wiley & Sons, Chichester.

Dixon, K. (1997) Gardening to conservation – the emerging role of botanic gardens in the recovery of endangered species. In D.H. Touchell and K.W. Dixon (eds) *Conservation into the 21st Century*. Proceedings of the 4th International Botanic Gardens Conservation Congress. Kings Park and Botanic Garden, West Perth, Western Australia, pp. 169–174.

Driver, B.L., Brown, P.J., Gregoire, T.G. and Stankey, G.H. (1987) The ROS planning system: evolution and basic concepts. *Leisure Sciences*, 9, 203–214.

Ewane Sumelong, J. (2001) The importance of community involvement in conservation. *Roots 23 Botanic Gardens Conservation International Education Review*, December 2001, 32–35.

Fennell, D. (2001) A content analysis of ecotourism definitions. *Current Issues in Tourism*, 4, 403–421.

Graham, R. (1992) VAMP: some summary notes. In E. Hamilton-Smith (ed.) *Park Visitor Research for Better Management*. Park Visitor Research Workshop, Phillip Institute of Technology 24–26 June 1992. Hepper Marriot and Associates, Victoria, pp. 13–42.

Grant Thornton Kessel Feinstein (GTKF) (2001) *Western Cape Tourism Trends Card*. Grant Thornton Kessel Feinstein and Western Cape Tourism Board, South Africa.

Guthrie, A. (1996) Conservation programmes at the Queen Elizabeth II Botanic Park. In P.S. Wyse Jackson and J. Willison (eds) *Plant Conservation in the Caribbean Islands: The Role of Botanic Gardens*. Proceedings of a Caribbean Islands Botanic Gardens Workshop, 29 June–1 July 1996, Holiday Inn, Grand Cayman. Botanic Gardens Conservation International, UK, pp. 13–15.

Hariramamurthi, G. (2000) The genesis of the Medicinal Plants Conservation Network (MPCN). *Roots 20 Botanic Gardens Conservation International Education Review*, July 2000, 33–36.

Hernández, M.R. (1992) Los 3 pasos a la educación. *Roots 5 Botanic Gardens Conservation International Education Review*, February 1992, 20–22.

Heywood, V. (1996) Botanic gardens and global change. In C. Hobson (ed.) *Botanic Gardens in a Changing World*. The Proceedings of the Third International Botanic Gardens Conservation Congress, Rio de Janeiro 19–25 October 1992. Botanic Gardens Conservation International, UK.

Higham, J. and Lück, M. (2002) Urban ecotourism: a contradiction in terms? *Journal of Ecotourism*, 1, 36–51.

Honey, M. (1999) *Ecotourism and Sustainable Development: Who Owns Paradise?* Island Press, Washington, DC.

Kohlmaier, G. and von Sartory, B. (1990) *Houses of Glass*. MIT Press, Cambridge, MA.

Lawton, L.J. and Weaver, D.B. (2001) Modified spaces. In D.B. Weaver (ed.) *The Encyclopedia of Ecotourism*. CABI Publishers, Wallingford, pp. 315–326.

Leiva Sánchez, A. (1996) Activities, situation and needs of national botanic gardens of Cuba, with some notes on plant conservation priorities in the country. In P.S. Wyse Jackson and J. Willison (eds) *Plant Conservation in the Caribbean Islands: The Role of Botanic Gardens*. Proceedings of a Caribbean Islands Botanic Gardens Workshop, 29 June–1 July 1996, Holiday Inn, Grand Cayman. Botanic Gardens Conservation International, UK, pp. 61–65.

Macnaghten, P. and Urry, J. (1998) *Contested Natures*. Sage, London.

Manfredo, M.J., Driver, B.L. and Tarrant, M.A. (1996) Measuring leisure motivation: a meta-analysis of the recreation experience preference scales. *Journal of Leisure Research*, 28, 188–213.

Mason, P. (2000) Zoo tourism: the need for more research. *Journal of Sustainable Tourism*, 8, 333–339.

Maunder, M. and Culham, A. (1997) The role of botanic gardens in the management of threatened island plants. In D.H. Touchell and K.W. Dixon (eds) *Conservation into the 21st Century.* Proceedings of the 4th International Botanic Gardens Conservation Congress. Kings Park and Botanic Garden, West Perth, Western Australia, pp. 207–221.

Michel, S. (1996) Establishing a network of botanic gardens in Haiti. In P.S. Wyse Jackson and J. Willison (eds) *Plant Conservation in the Caribbean Islands: The Role of Botanic Gardens*. Proceedings of a Caribbean Islands Botanic Gardens Workshop, 29 June–1 July 1996, Holiday Inn, Grand Cayman. Botanic Gardens Conservation International, UK, pp. 26–31.

Morris, K. (1997) Restoring the tradition: redefining the role of botanic gardens in East Africa. In D.H. Touchell and K.W. Dixon (eds) *Conservation into the 21st Century.* Proceedings of the 4th International Botanic Gardens Conservation Congress. Kings Park and Botanic Garden, West Perth, Western Australia, pp. 127–135.

National Botanical Institute (2000) *Annual Review 1999–2000*. National Botanical Institute, South Africa.

National Botanical Institute (2003) *Corporate Strategic Plan 2003–2006*. National Botanical Institute, South Africa.

Obua, J. (1997) The potential development and ecological impact of ecotourism in Kabale National Park, Uganda. *Journal of Environmental Management*, 50, 27–38.

Oikawa, J. (2000) *Future Role of Living Plant Collections in Gardens for Biodiversity Conservation*. Unpublished PhD thesis, University of Reading, UK.

Orams, M.B. (2001) Types of ecotourism. In D.B. Weaver (ed.) *The Encyclopedia of Ecotourism*. CABI Publishing, Wallingford, pp. 23–36.

Prest, J. (1981) *The Garden of Eden. The Botanic Garden and the Re-Creation of Paradise*. Yale University Press, New Haven and London.

Pushpangadan, P., Rajesekharan, S., George, V., Thomas, J., Koshy, K.C., Shanavaskhan, A.E. and Buni, S. (1998) Empowering the Community: Wealth for All. *Roots 17 Botanic Gardens Conservation International Education Review*, December 1998, 29–31.

Redclift, M. and Sage, C. (1994) Introduction. In M. Redclift and C. Sage (eds) *Strategies for Sustainable Development: Local Agendas for the Southern Hemisphere*. John Wiley & Sons, Chichester, pp. 1–16.

Ritchie, B.W. with Carr, N. and Cooper, C. (2003) *Managing Educational Tourism*. Channel View Publications, Clevedon.

Ryan, C. and Saward, J. (2004) The zoo as ecotourism attraction – visitor reactions, perceptions and management implications: the case of Hamilton Zoo, New Zealand. *Journal of Sustainable Tourism*, 12, 245–266.

Schreyer, R. and Beaulieu, J.T. (1986) Attribute preferences for wildland recreational settings. *Journal of Leisure Research*, 18, 231–247.

Schreyer, R. and Roggenbuck, J. (1978) The influence of experience expectations on crowding perceptions and social–psychological carrying capacities. *Leisure Sciences*, 1, 373–394.

Secretariat of the Convention on Biological Diversity (CBD) (2004) *Guidelines on Biodiversity and Tourism development: International guidelines for activities related to sustainable tourism development in vulnerable terrestrial, marine and coastal ecosystems and habitats of major important for biological diversity and protected areas, including fragile riparian and mountain ecosystems* (CBD Guidelines). Secretariat of the Convention on Biological Diversity, Montreal.

Seeni, S., Manickam, V.S., Sutherland, L. and Willison, J. (2002) People and plants: training Darwin mentors. *Roots 25 Botanic Gardens Conservation International Education Review*, December 2000, 33–39.

South African National Biodiversity Institute (2008) *History of Kirstenbosch: Missions and Mottos* [online]. Available at: http://www.sanbi.org/frames/kirstfram.htm [accessed: 7 March 2008].

Sutherland, L.A. (2000) *Ecotourism and Botanic Gardens*. Paper and workshop presented at the 1st World Botanic Gardens Congress, Asheville, NC, 26–30 June 2000.

Sutherland, L.A. (2005) *The Role of Botanic Gardens in Less Developed Countries in Nature-based Tourism and their contribution to Conservation and Sustainable Development*. Unpublished PhD thesis, London Metropolitan University, UK.

Symonds, A. (2000) Teacher capacity building: skills development through environmental education. *Roots 20 Botanic Gardens Conservation International Education Review*, July 2000, 25–28.

Symonds, A. (2001) Teacher capacity building: skills development through environmental education. In L.A. Sutherland, T.K. Abraham and J. Thomas (eds) *The Power for Change: Botanic Gardens as Centres of Excellence in Education for Sustainability*. Proceedings of the 4th International Congress on Education in Botanic Gardens, 8–12 November 1999. Tropical Botanic Gardens and Research Institute, Kerala, India, pp. 143–148.

The Economist (US) (2000) Travels in Cuba – the odd meal. *The Economist*, 4 March 2000, 87.

Tuan, Y. (1990) *Topophilia: A Study of Environmental Perception, Attitudes and Values*. Columbia University Press, New York, NY.

Wallace, B. (1997) Networking regional botanic gardens. In D.H. Touchell and K.W. Dixon (eds) *Conservation into the 21st Century*. Proceedings of the 4th International Botanic Gardens Conservation Congress. Kings Park and Botanic Garden, West Perth, Western Australia, pp. 45–48.

Waitt, G., Lane, R. and Head, L. (2003) The boundaries of nature tourism. *Annals of Tourism Research*, 30, 323–345.

Weaver, D. (2001) *Ecotourism*. John Wiley & Sons, Chichester.

Weaver, D. (2005) Comprehensive and minimalist dimensions of ecotourism. *Annals of Tourism Research*, 32, 439–455.

Wight, P.A. (1997) Ecotourism accommodation spectrum: does supply match the demand? *Tourism Management*, 18, 209–220.

Willison, J. (1997) Botanic gardens as agents for social change. In D.H. Touchell and K.W. Dixon (eds) *Conservation into the 21st Century*. Proceedings of the 4th International Botanic Gardens Conservation Congress. Kings Park and Botanic Garden, West Perth, Western Australia, pp. 339–344.

Wyse Jackson, P. and Sutherland, L. (2000) *International Agenda for Botanic Gardens in Conservation*. Botanic Gardens Conservation International, UK.

PART 3
The Future for Ecotourism and Environmental Sustainability

Chapter 13
Exploding the Myth of Ecotouri

Ken Simpson

> They took all the trees, put 'em in a tree museum …
> and they charged the people a dollar and a half just to see 'em
>
> (Joni Mitchell 1970)

Introduction

The 'tree museum' referred to in Joni Mitchell's *Big Yellow Taxi* (1970) is actually Honolulu's Foster Botanical Gardens, initially created by private owners in 1853 to showcase the more spectacular examples of flora that are indigenous to the Hawaiian Islands. Since being bequeathed to the city in 1930, the gardens have been extended from 5.5 acres to 13.5 acres, and their appeal has been enhanced through the addition of approximately 10,000 new species of rare and endangered plant life, many of which are *not* native to the Hawaiian Islands. In 2007, Foster Gardens welcomed more than 75,000 visitors, the majority of whom were resident outside of the state of Hawaii, and it now costs the people five dollars 'just to see 'em'. The curse of inflation!

Is this, then, a typical example of what we mean when we talk about ecotourism? There is a well-established school of thought that says that it could be, for initiatives such as these can easily be seen to align with the primary principles of ecotourism suggested by Dowling (1996). In this respect, Dowling notes that instances of ecotourism, as distinct from any other form of tourism, can be recognized through their demonstrable association with five core elements:

1. They are primarily based on nature.
2. They are ecologically sustainable.
3. They are environmentally educative.
4. They benefit the host community.
5. They generate tourist satisfaction.

On this basis, then, Foster Gardens is indeed an ecotourism site, and all of its visitors are thus ecotourists by both definition and default. And this form of 'ecotourism' is popular; similar sites have emerged at four other locations on the island of Oahu alone, and botanical garden visits have become a major element of Honolulu's vitally important inbound tourism industry. At first glance at least, the state and its residents can reasonably claim to have created an ecologically sustainable and environmentally educative resource, that is solidly based on elements of nature, and that has proven to be beneficial to host community and visitor alike. This sounds very much like a clear 'win–win' proposition that should be enthusiastically supported by all concerned.

Yet, perhaps this assessment is worth more than just a first glance. For example, would Foster Gardens look quite so compellingly attractive if we were to remove the ecotourism label, a legitimate question, as that label is a *prima facie* attribution of the author's for illustrative purposes only, and there is no intent to suggest that the Gardens themselves have ever attempted to present this perspective to their public. So, if we were to carry out an assessment of Foster Gardens as 'a living museum of tropical plants, some rare and endangered, which have been collected from throughout the world's tropics over a period of 150 years', which is what the gardens themselves claim to be (City and County of Honolulu undated), would we perhaps be inclined to adopt a more critical attitude?

Would we not wonder about the logic of a generic display of exotic (non-native) greenery in the heart of one of the planet's notably unique floral ecosystems? Would we not question the imposition of an entry fee to a facility that is typically provided free of charge by other civic administrations? Could we, therefore, not suggest that any attempt to classify gardens such as these, as 'ecotourism' attractions, might be little more than the application of marketing sleight-of-hand to a visitor site that has never aspired to anything so ambitious? It is this central proposition, that the apparently worthy concept of ecotourism has been largely hijacked by tourism industry operators in search of competitive advantage, that lies behind this chapter's contention that ecotourism may be much more beneficial in theory than it is in practice; and that both academic and practitioner interpretations of what the concept actually means have been seriously distorted by the triumph of economic considerations over ecological misgivings.

In support of this contention, the chapter first identifies the generally accepted foundations of the theoretical proposition that is referred to as ecotourism, before dissecting the reality of the tourist experience in terms of each of the foundation characteristics so identified. In this manner, a significant reality gap is identified, between the rose-coloured concept of ecotourism's professed sustainability and the rather less positive realities that tend to emerge at the coalface of tourism experience. A brief commentary is then offered on the extent to which this reality gap challenges our conventional thought processes, before the chapter concludes with the presentation of an alternative philosophical standpoint – a suggestion that both supply and demand sides of the tourism equation would be substantially better off if the word 'ecotourism' was eliminated from the tourism industry lexicon altogether.

Ecotourism Theory and Practice

In the half century since the introduction of larger capacity jet aircraft, increasingly affluent first world populations have progressively embraced regular holiday travel as an inalienable right rather than an isolated privilege, and one inevitable outcome of this apparently entrenched practice has been an enhanced desire and confidence to seek out places that are 'new and different'. More recently, though still dating back 20 years to the influential Brundtland Report into sustainable industrial development (WCED 1987), those populations have also begun to develop a more sophisticated understanding of the need for ecological conservation. For advocates and disciples of sustainable tourism development, this combination of increasing mobility, increasing curiosity and increasing sophistication has been generally recognized as providing the initial impetus for what would eventually become known as 'ecotourism'.

Although many of the 'new and different' places sought by growing numbers of travellers are inevitably located in second- and third-world environments that are ill-prepared for rapid increases in high-density visitor activity, Newsome et al. (2002) have observed that a uniformly positive outcome from tourism to those areas remains theoretically possible. For these authors, the costs and benefits of high tourism volumes are relatively easy to identify in advance of the event, and a carefully managed and multi-stakeholder pattern of response can succeed in the generation of significant levels of local economic benefit, without any apparent detriment to the environmental resource that simultaneously provides the motivation for visitor patronage and the setting for their post-arrival activities.

A central tenet of the ecotourism concept is, therefore, the proposition that, under an appropriate management regime, significant economic benefits can be generated with minimal (if any) accompanying costs. This chapter argues that this approach is excessively optimistic, and that the claimed emergence (and continued development) of a direct and symbiotic relationship between responsible and farsighted local communities on the one hand, and concerned and committed visitor groups on the other, is unrealistic in the extreme. In contrast, the chapter proposes that, like all forms of tourism activity, the implications of 'ecotourism' (assuming for the moment that there is in fact such a thing) will inevitably emerge from a complex blend of economic and social interactions between a destination with experiences to sell and a multi-faceted market with the resources to buy. As such, the level of sustainability that can reasonably be expected from any given form of tourism is a function of the interplay between all three parties to the equation; the destination environment, the character of participating experience providers, and the attitudes and behaviours of primary experience users. That level of sustainability cannot be assumed to be positive just because some individual or organization has attached the prefix 'eco-' to whatever experience is actually being enjoyed.

Yet this is often what appears to be happening. Repeated and apparently indiscriminate usage of the ecotourism label, when applied to a variety of tourism initiatives that share a professed commitment to sustainable development, has led to acknowledgement, by host and guest alike, of 'sustainable ecotourism' activities that in reality have little to do with either ecotourism or sustainability. In essence, continuous and consistent use of the word 'ecotourism' in the same breath as 'sustainable development' has led to an assumed association between the two concepts to such an extent that we have now largely accepted an unspoken assertion that 'if it is ecotourism, it must be sustainable'. Extreme examples of this phenomenon can be found in the analysis of both trophy hunting (Novelli et al. 2006) and traditional urban zoos (Ryan and Saward 2004) as potential candidates for inclusion as ecotourism activities.

Literature such as this leaves itself open to accusations of excessive attention to esoteric labelling and to commitment of excessive resources to the question of whether a given activity is or is not a legitimate example of ecotourism. As Harrison (1997: 75) succinctly puts it:

> In recent years ecotourism has become something of a buzzword in the tourism industry … academics have so busied themselves in trying to define it that they have produced dozens of definitions and little else.

In Search of Definition

As with many other elements of the study of tourism, ecotourism does indeed suffer from a lack of uncontested definition. The earliest recorded reference in the tourism literature is that of Ceballos-Lascurain (1980, cited in Boo 1990: xiv), who interpreted the concept as travel to:

> relatively undisturbed areas with the specific objective of studying, admiring, and enjoying the scenery and its wild plants and animals, as well as any existing cultural manifestations found in these areas.

This approach, which takes a visitor-centred perspective by representing ecotourism as a reason for travel rather than as a specific pattern of visitor behaviour, and by highlighting unspoilt destinations as an object of study rather than as a setting for discretionary leisure, has also provided the foundations for a definition first adopted by the World Tourism Organisation and subsequently modified by the original author. A generation on from his original thoughts, Ceballos-Lascurain was able to observe (1996: 20) that ecotourism could now be held to encapsulate:

> environmentally responsible, enlightening travel and visitation to relatively undisturbed natural areas in order to enjoy and appreciate nature (and any accompanying cultural features both past and present), that promotes conservation, has low visitor impact, and provides for beneficially active socio-economic involvement of local population.

Though the key elements of 'traveller motivation' and 'natural area' had been retained, there had been a decided emphasis shift away from visitor intent and towards visitor impact, away from culture as a subject of study and towards culture as part of a destination's profile. In short, the focus had swung away from tourists and their motivations, and towards the destination and its attributes.

The extent to which later literature has been able to reflect this movement may best be captured by the comment of Newsome et al. (2002) that ecotourism is insufficiently distinctive to justify its status as a discrete form of tourism by itself, and that it is more usefully perceived as an (admittedly important) subset of the larger field of 'tourism to natural environments'. According to these authors, a broad typology of natural tourism experiences exists, anchored by 'outdoors adventure' at the environmentally negative pole and 'ecotourism' at the benign or positive pole, with the primary goals of ecotourism reflecting a desire to foster sustainability through education in resource conservation, cultural revival and regional economic diversification. Again, this approach signals a further move towards the classification of ecotourism in terms of its consequences, rather than its origins, motives, or behaviours, a development exemplified by Martha Honey's 'seven characteristics of ecotourism' (1999) presented below:

1. Involves travel to natural destinations.
2. Builds environmental awareness.
3. Respects local culture.
4. Generates financial benefits for conservation.
5. Generates financial benefits and empowerment for local people.
6. Supports human rights and democratic movements.
7. Results in minimal impacts.

Honey's (1999) work provides a useful framework to guide any assessment of the value of ecotourism to a given destination. In this respect, if we were to examine any specific example of tourism activity, which involves the pursuit of specific experiences in a specific spatial environment, provided by specific operators to specific visitor groups, then Honey's criteria are clearly important in validating the overall costs and benefits of that particular instance of tourism development. Furthermore, if the example under review did, indeed, turn out to be largely positive when measured against *all* of Honey's criteria, then it would be reasonable to suppose that a uniformly beneficial form of ecotourism had in fact taken place. Unfortunately for all concerned, there appear to be few, if any, examples of this theoretically desirable outcome having been achieved in practice.

If this is the case, if the conditions implied by Honey's (1999) seven criteria are seldom, if ever, met in the real world of front line tourism operations, then surely the whole idea of ecotourism can be legitimately called into question. Certainly, conventionally imagined ecotourism activities clearly meet the first criterion, that of 'travel to natural destinations', but what about the others? If ecotourism does not build environmental awareness or respect local culture; if there are no financial benefits, either to enhance conservation or empower host communities; if, indeed, the impacts of ecotourism are indistinguishable from any other form of tourism, then surely we are justified in asking whether we should be undertaking this type of activity at all? How, then, does the ecotourism phenomenon, this 'travel to natural destinations', perform when measured against the remainder of Honey's (1999) criteria?

Environmental Awareness

This chapter was written in New Zealand, a country that is significantly isolated in terms of its physical location, and one that has therefore had to develop a uniquely compelling 'reason to visit' in order to compete for its share of international tourist expenditure. From a marketing perspective, it has succeeded brilliantly, initially with its adoption of a 'clean and green' market positioning, and then with the highly effective '100% Pure' strap line that is still used to brand all of the New Zealand Tourism Board's marketing communications. So, how does this marketing imagery translate into reality on the ground? Is it backed up by a deliberate strategy of sustainable development, effectively implemented through the co-ordinated activities of a sophisticated and responsible local tourism industry, or is it just that it is easy to disguise environmental failings in a country the size of Italy, with a population of four million people?

The reality is that, although there is an admirable element of highly responsible operating practice present in the way that New Zealand conducts its tourism relationships with the rest of the world, a significant number of unpalatable contradictions remain. The country has one of the developed world's highest rates of individual car ownership, with negligible controls on CO_2 emissions; one of the highest percentages of threatened species, in an environment in which many of these species are truly unique; seriously depleted fisheries which, in an island nation with an extensive economic exclusion zone, is arguably unforgivable; and a disturbingly high level of inland waterway pollution, primarily caused by the run-off from agricultural chemicals and pesticides.

The local tourism industry is aware of these contradictions, but there is little evidence that widespread knowledge has necessarily translated into greater care – the 'gorillas in the mist' idea that, when endangered resources are seen as more valuable when conserved than they would be when exploited, market forces alone are sufficient to generate positive outcomes. On a prominent 'ecotourism' lobby group website (Ecotours New Zealand undated), two local operators comment that, despite New Zealand being the last place on earth to 'feel the brunt of human occupation', it could be argued that nowhere on earth have so few people caused so much environmental destruction in such a short time. Meanwhile, on the same website, two central problems for New Zealand ecotourism are identified as 'low awareness of New Zealand's eco-attractions amongst international visitor groups' and 'limited marketing resources' to address that awareness deficiency.

In response, the Ecotours New Zealand website notes that ecotourism opportunities available to visitors include bird watching, whale and other marine mammal watching, and hiking/tramping, with visitors invited to 'enjoy dolphins, penguins, seals and albatross in their natural habitat, and to try trekking, kayaking and biking'. It does seem that the conventional definition of ecotourism is being a little stretched by some of those options, an interpretation that is supported by review of the recommended marketing strategy. Here, more effective use of the Internet is seen as 'the answer, (providing) exposure to millions of people … promoting New Zealand operators internationally … and inform(ing) international ecotourism travellers. By so doing, we hope to take a small step towards keeping New Zealand – clean and green'. It is difficult to follow this logic!

Respect for Local Culture

The impact of tourism on host community culture is a well documented theme in the literature, even to the extent of drawing a comment from Pope John Paul II that tourism in natural or under-populated areas can transform 'culture, religious ceremonies and ethnic festivities into consumer goods' (BBC News 2001). In some instances, such as Boyd and Butler's (1999) Northern Ontario case study, local people are themselves presented, both deliberately and involuntarily, as a significant component of local biodiversity, as another 'threatened species' that is worthy of examination, evaluation, categorization and conservation. These residents, especially indigenous residents, may then be more accurately visualized as a key component of the ecotourism parcel of visitor benefits, rather than as a primary and legitimate beneficiary of ecotourism outcomes.

As this type of relationship evolves, destination culture comes under increasing pressure to behave as visitors expect it to. Pseudo-cultural happenings are then created purely for the benefit of visitors, and family events such as church services, weddings and even funerals are routinely presented for tourist enjoyment. Ultimately, when local culture is deemed to be insufficiently interesting to offer the ubiquitous Kodak moment, there is always the possibility of inventing something. As Nicholson-Lord (2002) has noted, Thailand's traditional 'rooted in history' Chiang Mai flower festival was actually unknown in the city until the mid-1980s. The visitors wanted traditional festivals, so they got traditional festivals!

The clash between local community practices and visitor expectations can reach its zenith in circumstances where Western value systems may judge some aspects of local culture as ethically questionable or environmentally destructive. In this context, Buckley (2005) reports on the hunting and de-tusking of the narwhal by the Inuit of the High Arctic, commenting that this aspect of indigenous tradition continues in the face of a clear signal from the Canadian government that this practice is 'frowned upon'. Yet there was no attempt to disguise this activity from the tourist gaze of a party of visitors undertaking an 'ecotour' with the overt purpose of viewing these same marine mammals. As a result, an Inuit hunting expedition was seriously hampered by the presence of a party of 'outsiders', and the quality of the ecotourism experience was similarly reduced as the narwhal chose to stay well away from the tourists' motor launch on the logical assumption that the vessel contained Inuit hunters. Though these particular ecotourists had originally set out to witness the behaviour of an endangered animal in the wild, their eventual experience was more closely related to the observation of an indigenous community. As Buckley (2005) observes, however, a surprising number of 'ecotourists' were willing to buy narwhal tusks from the Inuit hunters!

Financial Benefits for Conservation

The International Ecotourism Society (undated) notes that the clearly defined demographic that is the international 'ecotourist' (well educated, well paid, and middle aged) has repeatedly expressed its willingness to pay for the privilege of participation in an authentic ecotourism experience. Yet, as Batta (2006) notes in his investigation into ecotourism in India's Himachal Pradesh province, the reality can be somewhat different. Batta (2006) comments that around 85 per cent of visitors to this mountainous region of Northern India come for scenic beauty, and have little knowledge of, or interest in, biodiversity or unique wildlife. In this particular instance, therefore, the 'eco-' prefix appears to owe its origins to a spurious marketing claim by the local tour operator community – there is little or no environmental interpretation available from a poorly trained tourism workforce, and the wider community appears largely ignorant of what tourism enterprises are doing in the region. Ecotourism in Himachal Pradesh is therefore virtually indistinguishable from mainstream visitor activity, especially in terms of its negative impacts, where facilities provided for visitors have resulted in the generation of large quantities of 'disposable' glasses, plates, cans and other packaging. When this is added to a significant degree of littering, deforestation, soil erosion, water scarcity and unscientific waste disposal, it seems likely that operators have made a conscious decision to accumulate social or community costs in preference to private costs of operation.

Turning his attention to the question of who should pay for a high level of environmental damage, Batta (2006) claims that almost half of all visitors were unwilling to pay any additional costs to compensate for environmental protection, and this reluctance to contribute was even more clearly established amongst local residents. A similar pattern was also apparent amongst operators, though attitudes appeared to be somewhat dependent on enterprise size, with resorts more willing to contribute than hotels, and hotels more positive than guest houses.

So, is the 'small is beautiful' platform actually a myth when it comes to calculating the costs and benefits of ecotourism? As with many related questions, the answer to this may be a cautious yes and no. While larger tourism businesses may be more acutely aware of business practices that are both philosophically and pragmatically desirable from an environmental protection viewpoint, this awareness comes at the cost of operational scale. For example, Peace (2005) cites the example of Kingfisher Bay Resort, on Australia's World Heritage listed Fraser Island, a high quality visitor accommodation venture that was consciously and overtly positioned as an 'eco-lodge' on its opening in 1992. Since then, the resort has continued to pioneer a number of innovative and creative ways to maintain its eco-image, and has been widely acclaimed as a benchmark for everything that is good about eco-lodge construction and operation. Yet Peace (2005: 323) notes that the resort's

> focal point is a central complex which includes reception areas, administrative offices, restaurants, bars, lounges and swimming pools. This complex is connected by two-tier wooden boardwalks to 152 hotel rooms. Then there is a semicircle of 110 self-contained villas as well as a shopping area (The Village) which incorporates a general store and petrol station, a bakery, a beauty parlour and a four wheel drive hire centre.

Though Kingfisher Bay does make a genuine and sincere effort to operate in a responsible manner, and though it makes a particular feature of educating its guests in the core principles of ecological sustainability, it remains a major man-made construction that is situated in an otherwise pristine wilderness location. Thus, though all possible steps have been taken to minimize negative impacts on the host environment, some of those impacts inevitably remain, not because the resort has been remiss or excessively tolerant, but because of the reality that *all* tourism infrastructure affects the location that supports it.

In reality, Kingfisher Bay is not an environmentally beneficial venture; it is a venture that has succeeded in keeping environmental costs to a minimum. Would it not therefore be better to abandon the 'eco' label altogether, and adopt a marketing profile that more accurately and more honestly depicts what the enterprise offers its guests – a comfortable oasis, in a remote natural area, that makes it easy for cosseted visitors to feel as if they are experiencing the great outdoors. Or would the abandonment of the 'eco-' prefix, as an implied guarantee of environmental beneficence, be too much of a price for the marketers to pay?

Financial Benefits and Empowerment for Local People

Batta's (2006) doubts about the financial benefits of ecotourism are especially marked when he considers the ultimate destination of tourist income in Himachal Pradesh. Echoing a concept that is fairly well established in the tourism impacts literature, he notes that just 30 per cent of total tourism revenue is currently being retained in the local area. Acknowledging that there is a significant economic outflow resulting from capital interest payments, repatriation of dividends to foreign owners, and imposition of central government taxation, the primary shortcomings are attributed to an inability and/or unwillingness of tourist operators to use local suppliers. In what is a predominantly agricultural community, 95 per cent of fruit and vegetables used by resorts, 65 per cent of similar products used by hotels, and 40 per cent of usage by guesthouses is sourced from outside the local area.

Yet still the tourists arrive, and still they expect to be cared for, although some writers have noted that a law of diminishing returns is beginning to make its presence felt. David Nicholson-Lord (2002) reports that, at the World Heritage site of Machu Picchu in Peru, local residents have rebelled against burgeoning tourist numbers by picketing the site of a cable car designed to take visitors to the attraction, and this extraordinary occurrence begs at least two salient questions. Firstly, how on earth has it become possible to construct an operating concept that permits inclusion of the words 'ecotourism' and 'cable car' in the same sentence; and, secondly, just how large is this volume of tourism that local people have decided they want a bigger slice of? At the time of writing (September 2007), the cable car was operating to capacity at 350,000 'ecotourism' visits annually. There appears to be little room for 'small is beautiful' in this example.

In reality, there are many such instances to support the claim that the financial proceeds of developing tourism do not normally accrue to individual members of the host community. Most of the inbound investment, and therefore the majority of output profit, is the preserve of larger airlines, transport operators and hoteliers. These companies, usually domiciled in a first world country, are openly dependent on a pristine local environment for much of the motivational force that generates individual leisure travel, but all too frequently regard the destination as little more than a golden egg-laying goose. As such, for eco-enterprises and conventional enterprises alike, low wages, predatory employment practices and high levels of import leakage are the norm. Additionally, as the destination grows and develops, the welfare of those who live there becomes increasingly irrelevant in the face of mounting opportunities for operator profit.

Support for Human Rights and Democratic Movements

As the historical cold war schism between East and West is replaced by an economic divide of even greater consequence, this time between North and South, the reality is that, in any form of tourism, there is an inevitable clash between the economic 'haves' of the Northern Hemisphere and their Southern counterpart 'have-nots'. Is ecotourism any less guilty than its mainstream counterparts in this respect? Probably not, though perhaps the well heeled, sophisticated, and respectfully middle-aged ecotourist may be a little more tactful, and a little less overtly disdainful, of his or her relationships with members of the host community.

Even in destinations where a late entry into the world of international tourism has enabled tourism planners to learn from the errors of others, one example being the belated participation of Cuba from the mid 1990s (Winson 2006), the most commonly adopted infrastructure remains the ubiquitous luxury hotel, owned and managed by foreign interests, located on fragile beachfronts, and using conventional and highly resource-hungry operating systems. In the face of the potential for large scale inbound investment, governments the world over have tended to display a low level of determination in protecting either individual human rights or democratic decision making.

In fact, as Belsky (1999) has noted, the emergence of ecotourism in second and third world countries can often precipitate the emergence of greater degrees of class, gender and other social distinction than existed prior to 'eco' arrivals. For example, James Mair (2006) reports an experience in the Las Perlas Archipelago in Panama where, across a two-week period in late 2006, the municipal authority had worked hard to establish a hydrological reserve to protect nearly 10,000 acres of wetlands that offered sanctuary to 14 endemic bird species; at the same time, the central government was announcing major large-scale tourist development plans for the same area. As Mair (2006) wryly comments, 'the left hand protects the ecology, the right hand builds on it'.

Minimal Impacts

As international travel continues to increase, with corresponding increases in both economic benefits and multi-faceted costs, the process of travel itself appears to be emerging as an easily identified and convenient villain of the piece. The issue of air miles consumed in visiting remote destinations, and the oxymoronic juxtaposition of travelling by air to help conserve the environment, is by now well enough established to negate the need for any further comment, other than to note two commonly overlooked issues. Firstly, aviation is the world's *fastest growing* source of greenhouse gas emissions; and, secondly, these gases are three times as harmful as ground-level gases due to the fragile composition of the high atmosphere (Hansford 2007).

Of equal concern, however, is the current posturing around carbon credit trading, a process which appears to have gained the support of many ecotourism devotees, but which this author prefers to portray as a mechanism by which one's environmentally destructive practices are set off against some form of compensating checks and balances. As a result, though it remains an environmental sin to fly long distances, the degree of sin is somehow reduced if the privilege is paid for. Surely this is facetious and self-delusory in the extreme, seeking to reassure our collective conscience that we have 'paid' for what will always be a discretionary leisure activity by engaging in some supposed package of compensating action.

The reality is that the worldwide volume of international tourism will almost certainly continue to increase, and ecotourism will by no means be immune from the consequences of this growth. In this respect, the International Ecotourism Society (undated) has itself noted that tourism to natural areas represented approximately 2 per cent of all leisure travel movements in the late 1980s, but that this proportion had grown to 20 per cent over the ensuing ten years. Based on these figures, it is possible that ecotourism in 2007 generates more than US$100 billion in annual visitor spending and that, again according to the Ecotourism Society, it continues to grow at a rate that is approximately three times faster than the overall industry total. Thus, and no matter what precautions are adopted, it is naïve to suppose that ecotourism impacts can be contained at or below current levels. Rather than maintain this essentially impracticable stance, it is much more relevant to consider the possibility that future ecotourism impacts may actually increase at a *greater* rate than other more conventional forms of tourist activity.

Where to From Here?

This chapter has attempted to illustrate a version of the 'power of brand', in respect of the clearly positive public image that proponents of ecotourism have managed to generate for a visitor activity that is little more than an (admittedly important) subset of tourism in natural areas. It has also attempted to make a case for the rejection of what Batta (2006: 42) calls 'the craze for calling any nature based tourism destination an ecotourism destination', on the basis that ecotourism is actually a *faux* concept whose defining characteristics are largely irrelevant to the much more important issues of tourism costs and benefits. As such, the more we try to define, interpret and apply the espoused principles of ecotourism activity, the greater is the danger that we will lose sight of the aspects of tourism development that really matter.

Yet the concept of ecotourism has a firm grasp on our collective psyche. We have an International Ecotourism Society, a dedicated *Journal of Ecotourism* and a veritable plethora of conferences for academics to attend. We have even experienced the ultimate acquisition of credentials through UN recognition of an International Year of Ecotourism in 2002. In addition, we know what our ecotourists look like, for a well-established demographic profile has been assembled, and this is an area for further research that will doubtless command a significant degree of support from both public and private sector tourism bodies. Such is the growth profile of this so-called industry sector that there is considerable incentive for tourism operators to maximize their attention, and commit their investment, to what may easily be seen as the next major cash cow.

Yet it may be that the concept of ecotourism is beginning to show signs of fraying around the edges, with Australian and New Zealand pavilions at this years ITB in Berlin being referred to as the 'Halls of the Devil' (Hansford 2007), a chilling endorsement of Nicholson-Lord's (2002) scathing attack on the concept:

> The reality (of ecotourism) is that no-one has properly defined it, no-one really knows what it means, it's highly unlikely that anyone will ever define it in a way that will command assent from critics of the industry, and in this vacuum the marketing men, greenwashers, corporate developers and government spin doctors flourish.

So, would it not be simpler to drop the pretence, to abandon the concept of ecotourism altogether, and to acknowledge that any tourism to natural areas will inevitably present a unique package of costs and benefits that reflect the political, societal and ecological realities of the specific circumstance under review? This has actually happened on at least one occasion, in Cuba in 2003, where the power of an authoritarian state allowed ecotourism to be formally declared a 'non-term', to be substituted instead with the broader concept of nature tourism (Winson 2006). Though cynics may readily perceive an ulterior motive, based on a governmental desire to directly control the evolution of tourism in that country, the stated intention is nevertheless sound – a focus on the characteristics of the destination as the most appropriate determinant of future development, and a belief that all tourism that takes place in the natural world must necessarily adopt ecologically sound attitudes towards the resource that it both relies on and seeks to exploit.

So let us reject the 'ecotourism' label as indicative of some imagined guarantee of ecological purity and innocence, a certification that the tourism experience concerned is in some way environmentally benign, educationally enlightening and locally focused. Let us instead concede that 'eco' sells and, as long as this remains the case, we will continue to be offered 'eco' shopping bags, golf tees, nappies and coffee mugs; we can continue to wear 'eco' shoes, eat 'eco' breakfast, clean our teeth with 'eco' toothbrushes and drain away the residue in an 'eco' InSinkErator – and, when all else fails, we can return to the earth in an 'eco' coffin that will biodegrade at a similar rate to its contents. Why then should we not be able to take 'eco' holidays?

The answer is, of course, that we can, and that we always will be able to, as long as someone, somewhere will be able to profit from us doing so. More important, however, is our continued ability to recognize 'eco' as something that works well as a central element in twenty-first century marketing communication, but is of considerably lesser effectiveness when viewed through any other eyes. The truly vital determinants of the future for natural area tourism do not lie with 'to 'eco', or not to 'eco''; rather they lie with the extent to which operators, and the communities they inhabit, are able to genuinely embrace a long-term view of what they are doing and what the consequences will be. As James Mair (2006) so memorably states:

> If only enough people can get into the symbiotic mindset of taking the long-term view instead of being part of the current prevailing short-term free-for-all 'grab and grow' culture, then might we save our children and grandchildren the mounting parasitic blight of our economic and environmental debts to them.

In each of the world's self-appointed ecotourism destination countries, there are a small number of dedicated and committed ecotourism operators who have already taken on board the vital necessity to embrace this perspective. However, there are a much larger number of pseudo-eco operations that have professed to do so while simultaneously going about their businesses according to a set of precepts that directly contradicts the ecotourism ethos. All we can hope for, given the accumulated wisdom of a generation of ecotourism operations, is the wisdom to be able to tell the difference between them.

References

Batta, R.N. (2006) Evaluating ecotourism in mountain areas: a study of three Himalayan destinations. *International Review for Environmental Strategies*, 6, 41–62.

BBC News (2001) *Mass Tourism: Is it Exploitation?* [online]. Available at: http://news.bbc.co.uk/1/low/talking_point/1403092.stm [accessed: 8 April 2008].

Belsky, J. (1999) Misrepresenting communities: the politics of community-based rural ecotourism in Gales Point Manatee, Belize. *Rural Sociology*, 64, 641–666.

Boo, E. (1990) *Ecotourism: The Potentials and Pitfalls*. World Wildlife Fund, Washington, DC.

Boyd, S.W. and Butler, R.W. (1999) Definitely not monkeys or parrots, probably deer and possibly moose: opportunities and realities of ecotourism in Northern Ontario. *Current Issues in Tourism*, 2, 123–137.

Buckley, R. (2005) In search of the narwhal: ethical dilemmas in ecotourism. *Journal of Ecotourism*, 4, 129–134.

Ceballos-Lascurain, H. (1996) *Tourism, Ecotourism, and Protected Areas*. World Conservation Union, Gland.

City and County of Honolulu (undated) *Foster Botanical Garden* [online]. Available at: http://www.honolulu.gov/parks/hbg/fbg.htm [accessed: 8 April 2008].

Dowling, R.K. (1996) The implementation of ecotourism in Australia. *Proceedings of 2nd International Conference of the Ecotourism Association of Australia, Bangkok, Thailand, 18–21 July*. Srinakharinwirot University, Bangkok.

Ecotours New Zealand (undated) Available at: http://www.ecotours.co.nz [accessed: 8 April 2008].

Hansford, D. (2007) Melting away. *New Zealand Listener*, 1 September, 28–31.

Harrison, D. (1997) Ecotourism in the South Pacific – the case of Fiji. In *World Ecotour '97 Abstracts Volume*. BIOSFERA, Rio de Janeiro.

Honey, M. (1999) *Ecotourism and Sustainable Development: Who Owns Paradise?* Island Press, Washington, DC..

Mair, J. (2006) *Ecotourism: A Sustainable Trade?* [online]. Available at: http://www.bbc.co.uk/caribbean/news/story/2006/12/061219_ecotourism.shtml [accessed: 8 April 2008].

Mitchell, J. (1970) *Big Yellow Taxi*. Alfred Publishing, Van Nuys, CA.

Newsome, D., Moore, S.A. and Dowling, R.K. (eds) (2002) *Natural Area Tourism: Ecology, Impacts and Management*. Channel View Publications, Clevedon.

Nicholson-Lord, D. (2002) *Green Tragedy: The Blight of Ecotourism* [online]. Available at: http://www.alternet.org/story/13371 [accessed: 8 April 2008].

Novelli, M., Barnes, J. and Humavindu, M. (2006) The other side of the ecotourism coin: consumptive tourism in Southern Africa. *Journal of Ecotourism*, 5, 62–79.

Peace, A. (2005) Managing the myth of ecotourism – a Queensland case study. *Australian Journal of Anthropology*, 16, 321–334.

Ryan, C. and Saward, J. (2004) The zoo as ecotourism attraction – visitor reactions, perceptions and management implications. *Journal of Sustainable Tourism*, 12, 245–266.

The International Ecotourism Society (undated) Available at: http://www.ecotourism.org/webmodules/webarticlesnet/templates/eco_template.aspx?a=12&z=25 [accessed: 8 April 2008].

WCED (World Commission on Environment and Development) (1987) *The Brundtland Report: Our Common Future*. Oxford University Press, Oxford.

Winson, A. (2006) Ecotourism and sustainability in Cuba – does socialism make a difference? *Journal of Sustainable Tourism*, 14, 6–23.

Chapter 14
Conclusion: Lessons Learnt and Ways Forward

Jennifer Hill and Tim Gale

Introduction

This book has attempted to offer conceptual and practical insights into the complex interactions between ecotourism and heterogeneous and dynamic natural environments. It has drawn on a diverse array of case studies from around the world, incorporating analysis from a range of geographic scales and integrating knowledge from the social and natural sciences. A variety of ecotourism activities, ecosystem types, ecosystem components and environmental responses have been examined, and it has been shown that, in order to be successful, ecotourism requires the maintenance of natural, socio-cultural, economic and political capital (Collins 1999, Scheyvens 1999, Kiss 2004, Jones 2005, Butcher 2006). The appeal of ecotourism, in enabling environmental conservation *and* socio-economic development (Boo 1990, Ross and Wall 1999, Cleaver 2001, Mohan 2002), rests in its potential to provide economic benefits whilst maintaining ecological integrity through sustainable resource consumption. From the thematic case studies presented in this volume, three overarching issues have emerged that may be summarized as: constructions of nature, ethics and environmental sustainability; ecotourism, nature and neoliberalism; and environmental measurement, monitoring and management. The general nature of these issues and their specific articulation within chapters is outlined here, before the book closes by proposing some new directions for ecotourism research.

Constructions of Nature, Ethics and Environmental Sustainability

There has been recent interest in the physical sciences concerning the emotional underpinning of scientific facts, particularly within the field of biogeography (see Trudgill 2001a, b, 2008, Fall 2008). Research in this area has shown that socio-cultural values mediate our response to scientific facts and, as such, whilst environmental conservation and ecotourism can be seen as rational projects (Adams 1997), our social constructs actually guide our individualistic understanding of them. In Chapter 2 of this volume, Andrew Holden introduces social constructions of nature, indicating how an anthropocentric view of nature assumes that it holds instrumental value for the purpose of human benefit. Tourism stakeholders adopting this view have a propensity to pursue resource destructive behaviour. Conversely, if nature is considered to possess a range of values beyond the instrumental, then tourism stakeholders may adopt more benign resource conserving behaviour. Such views influence individual perceptions of carrying capacity and levels of acceptable change in particular environments (see McCool and Lime 2001, Jennings 2004).

In Chapter 9, Leah Burns indicates that particular social constructions of wildlife can legitimize certain types of tourism management. She points out that as a consequence of an increase in tourist arrivals to Fraser Island in Queensland, Australia, interactions of people with dingoes have come to be classed progressively as negative. This has resulted in the identification, and subsequent culling, of dingoes as 'problem' animals (at a rate of approximately one per month). Such management, however, is set against the legal responsibility of the Queensland Parks and Wildlife Service to conserve dingoes as a protected 'native' species. The tension between anthropocentric values underpinning wildlife management and the positive construction of the dingo as a valuable part of contemporary Australian wildlife persists. Dingoes are variously seen as a pest versus a protected species, a native versus an introduced species and a pure-bred versus a hybrid species. Thus, the management approach to dingoes is related to how they are socially constructed. If they are recognized as 'protected, native and pure', then there is less incentive to cull them. If, however, they are considered as 'introduced, pest and hybrid', then culling becomes a more acceptable option in order to remove the possibility of contact with people (Hytten and Burns 2007a, b). Burns concludes that, rather than constructing humans as separate from nature, as reflected in current management strategies such as fencing, people should be viewed as part of nature, with management reconstructed to transfer a sense of responsibility to tourists. This may reduce the heavy burden of recreation for managers, *and* provide more satisfaction for tourists and a greater conservation value for wildlife.

In Chapter 12, Lucy Sutherland takes the idea of a socially constructed nature to the extreme with an examination of the potential role of, and challenges to, botanic gardens in promoting ecotourism within Less Developed Countries (LDCs). Her work complements the growing academic discourse on ecotourism in modified spaces (see, for example, Mason 2000, Lawton and Weaver 2001, Ryan and Saward 2004, Weaver 2005, Cater 2006, Weaver and Lawton 2007). Many botanic garden managers in LDCs are examining the role that their institutions can play in supporting the protection of local biodiversity and in building the capacity of local communities to use biodiversity sustainably, as well as generating much needed revenue to support their institutions' work in conservation and sustainable development. These operations complement the principles that ecotourism seeks to uphold. The case studies presented in Sutherland's chapter illustrate that some botanic gardens in LDCs already play a role in ecotourism, and further development of their role could alleviate pressure on protected areas by providing an alternative ecotourism experience. Moreover, the botanic garden experience can complement the protected area experience through education and awareness-raising programmes that prepare tourists for the latter, as well as addressing appropriate tourism behaviour in natural settings. The chapter concludes that, at a broader level, the socio-politics of the host country is a key factor hindering the integration of botanic gardens into ecotourism and, at a grass roots level, the passive approach taken by their management allows the tourism industry's Western view of botanic gardens in LDCs to persist (with the consequence that this particular construction of botanic gardens limits their recognition as viable settings for ecotourism).

Environmental sustainability is an inherently anthropogenic construct that can be examined at a number of spatial scales, including the global. In Chapter 3, Colin Hunter highlights the need to look beyond the immediate destination area when considering the environmental impact of tourism, particularly ecotourism with its frequent reliance on long-haul air travel. His findings suggest that ecotourism holidays involving international air travel are likely to produce an absolute demand on global natural resources. It is possible that even if a strong environmental conservation ethic exists locally at most ecotourism destinations, the cost to the environment of getting to these destinations will actually exert a significant impact on natural resource use and environmental change at the global scale. If ecotourism really is to act as a positive exemplar for the rest of the tourism industry, then there is a particular onus on those engaged in planning, delivering and promoting ecotourism products to consider their impact in more global terms, and to account for the demands placed on natural resources by air travel in environmental sustainability assessments.

Both Alton Byers (Chapter 4) and Wenjun Li (Chapter 5) highlight the importance of achieving conservation with equity within ecotourism by integrating local communities into particular ventures (see also Drumm 1998, Vincent and Thompson 2002, Kiss 2004, Cole 2006, Cusack and Dixon 2006). In Chapter 4, Byers notes that the most promising long-term conservation mechanisms to protect alpine environments in areas such as Nepal and Peru involve building local stakeholder capacity into the design, management and monitoring of projects. Likewise, in Chapter 5, Li presents a positive case study of tourism in the Jiuzhaigou Biosphere Reserve in China, where the environment has not been degraded extensively as tourism has developed, and the indigenous community and local region have benefited economically from the industry. The development and operation of a 'joint-stock company' to manage tourism, established between local community households and reserve administrators, is cited as a key element in the success of the project. The ceding of some authority and responsibility to local communities, allowing members to decide how local resources are used, has promoted local environmental stewardship and (more) sustainable tourism development.

In Chapter 8, Simon Evans considers the actual and potential roles of ecotourism in protecting vulnerable tiger populations in India, contextualising this within the broader ethical debate about 'fortress' conservation (as championed by conservation biologists) versus sustainable use (as championed by social ecologists). Whereas tiger conservation in India was initiated following the fortress conservation model (see Adams 1997, 2003, 2004, Brockington 2002, Hutton et al. 2005), with the establishment of reserves supported by policies to relocate local communities from inside protected boundaries, there have been recent calls for greater adherence to social utility, involving local people in conservation explicitly and rewarding them for the continued presence of wildlife within their neighbourhoods (see Hulme and Murphee 1999, Brown 2002, 2003, Adams et al. 2004, Brockington et al. 2006). Ecotourism can play a central role in such conservation, through provision of long-term economic incentives to protect threatened species. To achieve this, however, ecotourism must convey social benefits to local communities whilst following ethical practices on the ground. There is a danger that wildlife becomes manipulated in order to provide guaranteed sightings for tourists through, for example, baiting, which can lead to habituation (see also DeBruyn and Smith in Chapter 7 and Burns in Chapter 9). The marketing of ecotourism activities should be undertaken ethically, such that the expectations of tourists are kept realistic. Additionally, the income received from tiger tourism must be invested back into parks and their local populations. Evans concludes that, whilst tiger tourism represents a powerful tool with which to contribute to wider conservation efforts, at present this is not being realized. For ecotourism to contribute more effectively to tiger conservation efforts in India, an integrated strategy is required that views tourism development as a component of sustainable development, rather than as a sustainable option in its own right.

Two chapters in this volume provide evidence of ethics influencing the action of tourism stakeholders, notably through codes of conduct and wider visitor interpretation. In Chapter 10, ethical approaches to guiding responsible tourist behaviour are examined by Davina Stanford. She notes (as do Hill and Gough in Chapter 11) that ecotourism potentially draws tourists into fragile, remote and often protected areas that are vulnerable to the impacts of visitors. Such tourism needs to be controlled if the visitors are not to damage the places and wildlife that they are visiting. Control can be achieved through the provision of environmental information for tourists, increasing visitor knowledge and mindfulness (Moscardo 1996, 1999, Kimmel 1999). Stanford's research uses Kohlberg's stages of moral development as a framework to develop, and test empirically, several potential messages designed to influence seal-viewing behaviour at the Kaikoura Peninsula, New Zealand. The results highlight the importance of positive, credible and reasoned messages in influencing visitor behaviour; messages that appeal to good citizenship and emphasize the greatest good for the greatest number are most likely to influence visitor behaviour. By contrast, messages conveying universal ethical principles are least likely to influence visitor behaviour. Tourists in a novel situation do not intuitively know right from wrong and desired behaviour should, therefore, be explained to them in these messages, if they are to be effective.

Based on empirical research in Queensland, Australia, Jennifer Hill and Georgina Gough (Chapter 11) demonstrate that ecotourism interpretation can act not only to increase visitor satisfaction and knowledge, but also to reinforce visitors' already favourable conservation attitudes and behavioural intentions. They note the existence of a 'ceiling effect' (Beaumont 2001: 335), whereby visitors report reasonably strong environmental attitudes prior to their rainforest encounter that are little changed by immersion in the forest. Interpretive information does, however, seem to mediate the tourist encounter with nature, leading to the desire for tourists to gain a longer and more perceptive engagement with the environment in the future. Interpretation stimulates visitors to engage in a positive way with the rainforest, through recognition of personal responsibility for the protection of the environment and through the desire to teach others about such issues.

Finally, in Chapter 13, Ken Simpson identifies a significant reality gap between the ethics of ecotourism's professed sustainability and the rather less positive realities that tend to emerge at the coalface of tourism experience. He presents an alternative philosophical standpoint, suggesting that both supply and demand sides of the tourism equation would be substantially better off if the word 'ecotourism' was eliminated from the tourism industry lexicon altogether. Instead, he suggests it would be better to acknowledge that any tourism to natural areas will present a unique package of benefits and costs that reflect the political, societal and ecological realities of specific local contexts.

Ecotourism, Nature and Neoliberalism

The biophysical environment is becoming increasingly neoliberalized through mechanisms such as economic governance, valuation, privatization and enclosure (see, for example, McAfee 1999, McCarthy and Prudham 2004, Heynen and Robbins 2005, McCarthy 2005, Robertson 2006, Castree 2008a, b). Ecotourism can be said to contribute to this process (Duffy 2008), an issue that emerges strongly within the chapters of this volume. In Chapter 2, for example, Andrew Holden notes that a demand for ecotourism can provide impetus to commoditize (assign economic value to) natural resources, thereby prompting local and perhaps national desire for them to remain in a conserved state. The user-benefits, and hence economic value, of tourism are achieved from experiencing nature non-consumptively, maintaining it in a largely conserved state rather than exploiting it for alternative economic gain. By contrast, if natural habitats or wildlife are judged not to have sufficient economic value in comparison to other development options, then a pretext is established for their removal.

In Chapter 5, Wenjun Li notes that after withdrawal of government financial support for the Jiuzhaigou Biosphere Reserve in the early 1990s, the natural resources in the reserve were entered into the market economy. Reserve administrators decided upon a 'joint-stock mechanism' with local residents as the operative system, partly because such systems were encouraged by central government and were widely used in China's state enterprises. The adoption of this mechanism in the reserve, therefore, was embedded in the context of national economic policy. Li concludes that the system has been generally positive: it has kept environmental changes within an acceptable level and has allowed local communities to benefit from tourism.

The power of the environment as a strategic commodity within neoliberal market economies has caused Ken Simpson (Chapter 13) to question whether ecotourism has been largely hijacked by tourism industry operators, who adopt the 'eco' label spuriously in order to gain competitive advantage. He offers examples of poor socio-environmental practices on the ground, despite such product branding. This issue is also highlighted by Simon Evans (Chapter 8), who notes that the perceived performance of ecotourism in tiger reserves in India has been compromised by entry into the market of a number of tourism enterprises employing the 'eco-' prefix as a marketing label, without ever having to justify their performance in relation to conservation or the pursuit of sustainability. Thus, sustainability cannot be assumed just because the experience being sold has been branded in this fashion. There is, indeed, an important role for tourists – if they are capable of it – to discriminate between products and, in turn, offer competitive advantage to truly dedicated ecotourism operators.

The neoliberalization of nature, which sees the market become the optimum mechanism for allocating environmental goods and services to meet the diverse needs of actors across the globe (Castree 2008a), may be perceived as a positive or negative process. It has afforded greater agency to the environment and wildlife in the tourism–environment relationship, allowing the entry of nature in a conserved state into global markets and promoting the sustainable use of the biophysical environment as a competitive form of landuse. Consequently, it can be said to form part of 'new conservation', notably through the integration of community natural resource use with environmental protection, mediated through local understanding of ecosystem structure and functioning (Hulme and Murphee 1999, Brown 2002, 2003). This, in turn, might help to heal the currently fractured relationships between the conservation, tourism and development sectors. However, there remains the difficulty of reconciling multiple perspectives concerning the value of nature. How are these value differences catered for within a market economy? What if environmental values and services exceed those registered through market prices? Additionally, what is the value of ethics within such a financial system? In seeking answers to these questions, the market may actually be viewed as a contentious way of distributing environmental services among multiple stakeholders with differing socio-economic backgrounds.

Environmental Measurement, Monitoring and Management

Any debate concerning the inter-relationships between ecotourism and the environment requires some ability to measure past, present and future states of different environments and their value within market economies as perceived by different stakeholders. In Chapter 4, Alton Byers stresses the importance of employing multiple methods of environmental measurement in order to monitor sustainability in the alpine zones of Sagarmatha National Park, Nepal and the Huascarán National Park, Peru. Long-term monitoring, using methods from the social and natural sciences (also advocated by Li in Chapter 5), has highlighted that alpine zones within both parks have been significantly degraded during the past 25 to 40 years as a result of poorly controlled adventure tourism, particularly trekking and mountaineering. Impacts within the alpine zone have included the over-harvesting of fragile alpine shrubs and plants to provide fuel for expeditions and tourist lodges, overgrazing, accelerated hillslope erosion, uncontrolled lodge building, health hazards related to poor sanitary practices and improper garbage disposal. Byers notes that more interdisciplinary and applied research is essential for the better understanding of differing threats to alpine integrity, for remedial interventions and for capacity-building within local user groups. He notes that the Alpine Conservation Partnership project, with promising beginnings in the Himalayan and Andes mountains, represents the first global initiative to actively protect and restore alpine ecosystems, whilst building the management capacities of the people who live there.

Efforts to value biophysical resources in economic terms are of fundamental relevance to the process of neoliberalising nature. In Chapter 2, however, Andrew Holden indicates that comparative measures of the economic value of natural resources in a conserved state are limited (see, also, Boo 1990). It is consequently difficult to argue for the continued existence of the environment in a conserved state vis-à-vis its modification for alternative economic activities. In Chapter 8, Simon Evans notes a further problem, originating in the range of conflicting definitions of ecotourism (see Fennell 2001). This causes difficulties in assessing what ecotourism actually is and how it should be measured. Depending on the definition employed, the economic value of ecotourism generates anything from US$30 billion (Honey 1999) to US$1.2 trillion (Ceballos–Lascurain 1996).

In Chapter 3, Colin Hunter argues that, in order to test the environmental impacts of ecotourism at the global scale, a metric is required that can express different types of impact generated at different stages in the life cycle of a tourism product. He offers Ecological Footprint analysis as a means of identifying and understanding globally expressed demands on the biosphere brought about by tourism activity, whilst also noting that caution is required in the interpretation of figures gained from such analyses, partly due to the use of proxy measures. It is only through the collection of primary data for real ecotourism products that assumptions can be tested rigorously. He notes that approaches to data collection might include the keeping of diaries by tourists, visitor surveys, and gathering information from hotels, restaurants and the providers of excursions. He also comments that a promising avenue for future research may be to utilize an input–output macroeconomic framework to redistribute and disaggregate national Ecological Footprint analysis information down to the household and tourist economy levels, using expenditure data (Wiedmann et al. 2005).

A number of chapters in the book make reference to the need to improve the planning and management of ecotourism, particularly through regionally–coordinated initiatives informed by academic research. In Chapter 6, Emma Stewart and Dianne Draper suggest that Arctic cruise tourism has moved beyond its infancy and is now entering a phase of maturity, with increased numbers of vessels and more regular patterns of activity. Despite this growth, these authors note that there has been little coordinated, trans-regional planning for the sustained development of cruise tourism in Arctic Canada, partly due to the complex involvement and interaction of multi-scale stakeholders in this region. Mindful of the inherent difficulties of collaborative practice, Stewart and Draper propose that cruise operators, communities and tourists might benefit from cruise tourism planning being given higher priority in the broader framework of integrated coastal management. Such management translates into a number of potential initiatives: including communities in establishing the upper limits of passenger numbers who disembark (thereby determining destination carrying capacity); in deciding for themselves how to present their culture to visitors; in developing particular interpretive approaches to ensure that pre-determined messages are communicated effectively to visitors; and in involving members in policy development. Similarly, tour operators could cooperate to try and achieve a greater distribution of economic benefits among Arctic communities, whilst distributing tourists more widely across the environment; they could also implement voluntary codes of conduct, thereby promoting acceptable behaviour amongst their passengers when interacting with local communities. Finally, the tourists could adhere to voluntary codes of conduct; contribute financially to each community visited; educate themselves about the places they visit before they disembark; and engage in ambassadorial activities.

In Chapter 7, Terry DeBruyn and Tom Smith explore the management of bear-viewing in Alaska. A growing demand for this activity in the State presents a challenge to resource managers as they strive to minimize human impacts on both the bears and associated wilderness. Areas that concentrate bears attract the greatest numbers of visitors, and these areas are nearly always critical feeding sites. Hence, the potential for negative impacts on bears by viewers is considerable. Habituation to human disturbance is maladaptive for bears and, as a result, viewing programmes in some areas have led to the displacement of individual animals, particularly adult males. As few studies have examined the interactions between bear-viewing and bear behaviour, the full extent of any behavioural or resource use change as regards the species is unknown. DeBruyn and Smith conclude that it is essential for agencies to develop a unified regional approach to bear-viewing policies and programmes within Alaska, specific to each bear-viewing area, which provide for optimal visitor experiences and safety whilst minimizing their impact on bears. Long-term monitoring can then determine whether management actions are effective in achieving desired outcomes, and any necessary adjustments can be made through adaptive management.

Finally, in Chapter 11, Jennifer Hill and Georgina Gough note that environmental interpretation and visitor education might offer a ‘soft’ strategy for managing tourism that complements many 'hard' physical and regulatory options, thereby allowing a more holistic approach to management (Kuo 2002, Mason 2005, Marion and Reed 2007). However, whether such environmental education can support a shift from a 'minimalist mode' of ecotourism (superficial learning opportunities and site-specific sustainability objectives) to a 'comprehensive mode' (fostering deep understanding and transformation of behaviour) remains to be seen (see Weaver 2005, for more information about these terms).

New Directions in Ecotourism Research

We would not presume to dictate an agenda for ecotourism (and environmental sustainability) research. However, there are opportunities for new knowledge creation associated with alternative ways of seeing the subject(s) of this book. Firstly, we believe that ecotourism research can be refreshed by ideas and techniques drawn from Mobilities, a new 'movement driven' social science incorporating the work of geographers and sociologists (Urry 2007). This has certain outcomes for researching tourism in general, which may be summarized as follows:

1. A movement towards 'de-exoticising tourism', so that we are no longer preoccupied with tourists and travels to distant lands.
2. Recognition that the processes which have enhanced the mobility of some people(s) have also served to highlight, and to heighten, the immobility of others.
3. The inclusion of imaginative, virtual and communicative (as well as physical or corporeal) travel in analyses of tourism phenomena, which are made possible by a range of digital devices and networks that allow us to access other people and places at the touch of a button.
4. Concern for the undesirable and hitherto unforeseen consequences of living a mobile life that lie outside of our control, and which threaten that very mobility (e.g. climate change, terrorist atrocities).

The implications for tourism studies of the so-called 'mobility turn' in the social sciences are further explored in Gale (2008, 2009). The concern, here, is what this means for the sub-field of ecotourism. Looking back at earlier chapters it can be seen that ecotourism is not simply what ecotourists, or those that cater to their needs and wants, do. Rather, it encompasses the multiple and intersecting mobilities of various individuals (for example, hosts, guests and those not directly implicated in the tourism system), and of other sentient and non-sentient life forms, buildings, objects and machines. Likewise, ecotourism sites and destinations are not fixed and given, but arise from the contingent conjunction of these agents at particular points in time–space, in order to realize certain performances that are recognisable as 'ecotourism'. Thus, for example, the patterns of movement exhibited by members of the brown bear or wild tiger populations of Alaska and

India respectively, as a consequence of habituation to humans through ecotourism (see Chapters 7 and 8), are no less significant than the motivations and behaviour of the 'ecotourists' themselves (traditionally the focus of much ecotourism research). Elsewhere in the book, attention has been drawn to the comparative *immobility* of certain groups of people (for example, the communities located on the routes to and from the reserves mentioned in Chapters 5 and 8, or those that play host to visitors disembarking from cruise ships in the Canadian Arctic, as discussed in Chapter 6), and to the reciprocal relationship between climate change and (eco)tourism as both 'perpetrator' and 'potential victim', given its dependency on long-distance travel and vulnerable natural resources (see Chapters 2 and 3). In addition, we might infer from the above list of outcomes that ecotourism and the benefits thereof are not necessarily confined to remote locations and are capable, in fact, of being experienced 'closer to home', whether this be in respect of using new information and communication technologies to create surrogate experiences in modified settings within or adjacent to urban centres, such as zoos and botanic gardens (see Chapter 12 for a discussion of the latter), or with regards to the significance attached (in Chapter 11) to interpretation provision at natural attractions in promoting positive attitudes and behaviour change towards the environment, both at the destination *and* following one's reintegration into everyday life (such as recycling household waste). No doubt, readers will draw further conclusions from our assertion that ecotourism is best seen as a form of temporary mobility, and we invite them to do so.

Secondly, and with the benefit of hindsight, we would question the efficacy of the term 'environmental sustainability', substituting it with 'environmental justice' (inspired by Lee and Jamal 2008). Turning this on its head, there are many *injustices* manifest in the biophysical and social environments, when using ecotourism as an incentive to promote improved environmental management; these include 'collateral damage' to flora and fauna when tracking certain species such as the 'Big Five' in southeast Africa (lions, elephants, buffaloes, leopards and rhinoceros), 'anti-social' behaviour in animals (as attributed to the dingoes of Fraser Island, in Chapter 9) resulting from stress levels being exceeded, and the reduction or loss of access to natural resources by the rural poor due to the perception that ecotourists do not want to see local people in national parks and other protected areas (Lewis 2008; see also Chapter 8). Furthermore, the pursuit of environmental justice in and through ecotourism, though ambitious, might be more attainable in the medium- to long-term, and easier to measure, when compared to an objective as open to interpretation (and misappropriation) as 'environmental sustainability'. Of course, some might dismiss this as mere semantics, but if ecotourism – or indeed any form of tourism – is managed in such a way as to maximize the chances of it being 'just' or 'fair' (so that it does not degrade the environment, and counters forces that marginalize and impoverish the most vulnerable in society), then it is also likely to be (more) sustainable. Hence, just as the term 'ecotourism' has been (mis)used so often as to become virtually redundant (see Chapter 13), so might we wish to mark the beginning of the end of '(environmental) sustainability' as a buzzword.

Finally, by way of a footnote to this edited volume, we would remind readers that the enablements and constraints to using ecotourism as a tool for promoting environmental sustainability (or environmental justice, if one agrees with the above recommendation) are related to complex and overlapping processes at various scales, from the local to the global, which produce geographically distinct or contextualized outcomes that one cannot assume to be transferable to other contexts. Hence, there are no claims as such to good practice (even if some of the initiatives and techniques described in the book, for example in Chapters 4 and 10, inspire good practice elsewhere). Accordingly, we would hold that there is a place for research about ecotourism (and environmental sustainability) that offers few answers but raises many questions.

References

Adams, W.M. (1997) Rationalization and conservation: ecology and the management of nature in the United Kingdom. *Transactions of the Institute of British Geographers*, 22, 277–291.

Adams, W.M. (2003) *Future Nature: A Vision for Conservation* (2nd edition). Earthscan, London.

Adams, W.M. (2004) *Against Extinction: The Story of Conservation*. Earthscan, London.

Adams, W.M., Aveling, R., Brockington, D., Dickson, B., Elliot, J., Mutton, J., Roe, D., Vira, B. and Wolmer W. (2004) Biodiversity conservation and the eradication of poverty. *Science*, 306, 1146–1149.

Beaumont, N. (2001) Ecotourism and the conservation ethic: recruiting the uninitiated or preaching to the converted? *Journal of Sustainable Tourism*, 9, 317–341.

Boo, E. (1990) *Ecotourism: The Potentials and the Pitfalls* (Vol. 1). World Wildlife Fund, Washington, DC.

Brockington, D. (2002) *Fortress Conservation: The Preservation of the Mkomazi Game Reserve*. James Currey, Oxford.

Brockington, D., Igoe, J. and Schmidt–Soltau, K. (2006) Conservation, human rights and poverty reduction. *Conservation Biology*, 20, 250–252.

Brown, K. (2002) Innovations for conservation and development. *The Geographical Journal*, 168, 6–17.

Brown, K. (2003) Three challenges for real people-centred conservation. *Global Ecology and Biogeography Letters*, 12, 89–92.

Butcher, J. (2006) Natural capital and the advocacy of ecotourism as sustainable development. *Journal of Sustainable Tourism*, 14, 529–544.

Castree, N. (2008a) Neoliberalising nature: the logics of deregulation and reregulation. *Environment and Planning A*, 40, 131–152.

Castree, N. (2008b) Neoliberalising nature: processes, effects, and evaluations. *Environment and Planning A*, 40, 153–173.

Cater, E. (2006) Ecotourism as a Western construct. *Journal of Ecotourism*, 5, 23–39.

Ceballos-Lascurain, H. (1996) *Tourism, Ecotourism and Protected Areas*. IUCN, Gland.

Cleaver, F. (2001) Institutions, agency and the limitations of participatory approaches to development. In B. Cooke and U. Kothari (eds) *Participation: The New Tyranny?* Zed Books, London, pp. 36–55.

Cole, S. (2006) Information and empowerment: the keys to achieving sustainable tourism. *Journal of Sustainable Tourism*, 14, 629–644.

Collins, A. (1999) Tourism development and natural capital. *Annals of Tourism Research*, 26, 98–109.

Cusack, D. and Dixon, L. (2006) Community-based ecotourism and sustainability: cases in Bocas del Toro Province, Panama and Talamanca, Costa Rica. *Journal of Sustainable Forestry*, 22, 157–182.

Drumm, A. (1998) New approaches to community-based ecotourism management. In K. Lindberg, M.E. Wood and D. Engeldrum (eds) *Ecotourism: A Guide for Planners and Managers* (Vol. 2). The Ecotourism Society, North Bennington, Vermont, pp. 197–213.

Duffy, R. (2008) Neoliberalising nature: global networks and ecotourism development in Madagascar. *Journal of Sustainable Tourism*, 16, 327–344.

Fall, J. (2008) *Global swarming: cosmopolitan natures, 'alien' species and the politics of barbarian plants*. Paper presented at the RGS–IBG Annual Conference, Kensington Gore, London, 27–29 August.

Fennell, D. (2001) A content analysis of ecotourism definitions. *Current Issues in Tourism*, 4, 403–421.

Gale, T. (2008) The end of tourism, or endings in tourism? In P. Burns and M. Novelli (eds) *Tourism and Mobilities: Local–Global Connections*. CABI Publishing, Wallingford, pp. 1–14.

Gale, T. (2009) Urban beaches, virtual worlds and 'the end of tourism'. *Mobilities*, 4, 119–138

Heynen N. and Robbins, P. (2005) The neoliberalization of nature: governance, privatisation, enclosure and valuation. *Capitalism, Nature, Socialism*, 16, 1–4.

Honey, M. (1999) *Ecotourism and Sustainable Development: Who Owns Paradise?* Island Press, Washington, DC.

Hulme, D. and Murphee, M. (1999) Communities, wildlife and the 'new conservation' in Africa. *Journal of International Development*, 11, 277–285.

Hutton, J., Adams, W.M. and Murombedzi, J.C. (2005) Back to the barriers? Changing narratives in biodiversity conservation. *Forum for Development Studies*, 32, 341–370.

Hytten, K. and Burns, G.L. (2007a) *Dingo Dualisms: Constructions of Nature and Culture in the Australian Context*. Paper given at the Animals and Society Conference, Tasmania, July 2007.

Hytten, K. and Burns, G.L. (2007b) Deconstructing dingo management on Fraser Island, Queensland: The significance of social constructionism for effective wildlife management. *Australasian Journal of Environmental Management*, 14, 40–49.

Jennings, S. (2004) Landscape sensitivity and tourism development. *Journal of Sustainable Tourism*, 12, 271–288.

Jones, S. (2005) Community-based ecotourism: the signifincace of social capital. *Annals of Tourism Research*, 32, 303–324.

Kimmel, J.R. (1999) Ecotourism as environmental learning. *Journal of Environmental Education*, 30, 40–44.

Kiss, A. (2004) Is community-based ecotourism a good use of biodiversity conservation funds? *Trends in Ecology and Evolution*, 19, 232–237.

Kuo, I.–L. (2002) The effectiveness of environmental interpretation at resource–sensitive tourism destinations. *International Journal of Tourism Research*, 4, 87–101.

Lawton, L.J. and Weaver, D.B. (2001) Modified spaces. In D.B. Weaver (ed.) *The Encyclopedia of Ecotourism*. CABI Publishing, Wallingford, pp. 315–326.

Lee, S. and Jamal, T. (2008) Environmental justice and environmental equity in tourism: missing links to sustainability. *Journal of Ecotourism*, 7, 44–67.

Lewis, F. (2008) *Handing out the justice? Tourism: environmental incentive or injustice?* Paper presented at the CeLTS–ESRC Seminar Series: Tourism, Inequality and Social Justice: Environmental, Economic and Social Futures for all? University of the West of England, Bristol, 24 September.

Marion, J.L. and Reed, S.E. (2007) Minimizing visitor impacts to protected areas: the efficacy of low impact education programmes. *Journal of Sustainable Tourism*, 15, 5–27.

Mason, P. (2000) Zoo tourism: the need for more research. *Journal of Sustainable Tourism*, 8, 333–339.

Mason, P. (2005) Visitor management in protected areas: from 'hard' to 'soft' approaches? *Current Issues in Tourism*, 8, 181–194.

McAfee, K. (1999) Selling nature to save it? Biodiversity and green developmentalism. *Environment and Planning D: Society and Space*, 17, 133–154.

McCarthy, J. (2005) Devolution in the woods: community forestry as hybrid neoliberalism. *Environment and Planning A*, 37, 995–1014.

McCarthy, J. and Prudham, S. (2004) Neoliberal nature and the nature of neoliberalism. *Geoforum*, 35, 275–283.

McCool, S.F. and Lime, D.W. (2001) Tourism carrying capacity: tempting fantasy or useful reality? *Journal of Sustainable Tourism*, 9, 372–388.

Mohan, G. (2002) Participatory development. In V. Desai and R. Potter (eds) *The Companion to Development Studies*. London, Edward Arnold, pp. 49–53.

Moscardo, G. (1996) Mindful visitors: heritage and tourism. *Annals of Tourism Research*, 23, 376–397.

Moscardo, G. (1999) *Making Visitors Mindful*. Sagamore, Champaign, IL.

Robertson, M. (2006) The nature that capital can see: science, state, and market in the commodification of ecosystem services. *Environmental and Planning D: Society and Space*, 24, 367–387.

Ross, S. and Wall, G. (1999) Ecotourism: towards congruence between theory and practice. *Tourism Management*, 20, 123–132.

Ryan, C. and Saward, J. (2004) The zoo as ecotourism attraction – visitor reactions, perceptions and management implications: the case of Hamilton Zoo, New Zealand. *Journal of Sustainable Tourism*, 12, 245–266.

Scheyvens, R. (1999) Ecotourism and the empowerment of local communities. *Tourism Management*, 20, 245–249.

Trudgill, S.T. (2001a) Psychobiogeography: meanings of nature and motivations for a democratized conservation ethic. *Journal of Biogeography*, 28, 677–698.

Trudgill, S.T. (2001b) *The Terrestrial Biosphere: Environmental Change, Ecosystem Science, Attitudes and Values*. Pearson, Harlow.

Trudgill, S.T. (2008) A requiem for the British flora? Emotional biogeographies and environmental change. *Area*, 40, 99–107.

Urry, J. (2007) *Mobilities*. Polity, Cambridge.

Vincent, V.C. and Thompson, W. (2002) Assessing community support and sustainability for ecotourism development. *Journal of Travel Research*, 41, 153–160.

Weaver, D. (2005) Comprehensive and minimalist dimensions of ecotourism. *Annals of Tourism Research*, 32, 439–455.

Weaver, D.B. and Lawton, L.J. (2007) Twenty years on: the state of contemporary ecotourism research. *Tourism Management*, 28, 1168–1179.

Wiedmann, T., Minx, J., Barrett, J. and Wackernagel, M. (2005) Allocating ecological footprints to final consumption categories with input–output analysis. *Ecological Economics*, 56, 28–48.

Index